MATH *for* CARPENTRY *and* CONSTRUCTION

RICHARD B. MILES

Second Edition

Publisher
The Goodheart-Willcox Company, Inc.
Tinley Park, IL
www.g-w.com

Library of Congress Control Number: 2022934699

ISBN 978-1-63776-706-1

1 2 3 4 5 6 7 8 9 – 24 – 27 26 25 24 23 22

Image Credits. Front cover: ungvar/Shutterstock.com (circular saw), cryptographer/Shutterstock.com (ruler); Section opener images: cryptographer/Shutterstock.com; Unit opener images: ungvar/Shutterstock.com

Preface

Math for Carpentry and Construction was developed to bridge the abstract principles taught in academic math to the real-world problems a carpenter must solve in the construction trades. With today's emphasis on standardized testing in education, this book was developed as a tool to help you become proficient in accurately performing trade-related mathematical problems.

This second edition of ***Math for Carpentry and Construction*** contains two new resources. The first is the addition of Section 6, *Material Estimating Activities*. Every carpenter needs to know how to estimate a job to give a quote to a customer, or to order materials for an upcoming job. This final section consists of 10 activities that will teach students the formulas and steps used to estimate materials. These activities provide a great opportunity to draw upon the math skills they have learned in the text. The second is a new resource entitled Appendix A, *Construction Diagrams and Terms*. This additional Appendix will assist the student when working through this text and will be an exceptional resource during their career.

After 16 years of working in residential, modular, and commercial construction, I entered secondary education with the goal of teaching high school students the trade of carpentry and construction. I was shocked at how many students could not measure or apply basic mathematical concepts to construction trade problems. I quickly realized that before I could properly teach my trade, I first needed to raise the competency level in mathematic concepts, and then apply them to construction principles. With over 25 years of experience in the classroom laying a solid foundation of applied math concepts, I have assembled those principles in ***Math for Carpentry and Construction*** to help instructors lay that same solid math foundation in their classrooms.

Based on my observation and experience, academic mathematic instruction has been driven by a culture of standardized testing. Because of this, students are not always learning the concepts necessary to be successful in a trade career, only the skills necessary to pass a standardized test. Often while teaching a concept the students have already covered in math class, I can see that moment when things begin to make sense because I have shown them how to apply it in a trade application. ***Math for Carpentry and Construction*** has been developed first and foremost with the purpose of teaching you the skills necessary and to make those connections so you will be proficient in the construction trades. This ensures that the student will be prepared for national and trade certification exams and a successful career in the construction trades.

All the problems in ***Math for Carpentry and Construction*** can be performed with the use of a calculator. Because it is my belief that every student should know how to work the problem without a calculator, every concept has been explained step-by-step so you can learn the concepts and process to complete any problem you will face in the field by hand. A calculator will give you the right answer, but you will not develop an understanding of how the answer was achieved. This will not help you the day you need to perform the math and do not have an electronic device.

Math for Carpentry and Construction is the best tool on the market today to learn math concepts used in the construction industry. It covers mathematic principles in a logical, applied manner, so that you can become mathematically proficient in your career.

Richard B. Miles

About the Author

Richard B. Miles is an instructor at Columbia-Montour Area Vocational-Technical School where he teaches residential construction. Mr. Miles earned his bachelor's degree in workforce education and development from Pennsylvania State University and his Trade Competency and Vocational II certifications in carpentry. He taught carpentry and career education at a residential treatment center and has served as a subject-matter expert designing and editing trade exams for the National Occupational Competency Testing Institute (NOCTI). Mr. Miles' work experience includes over 25 years in secondary education teaching carpentry and construction and career education. Mr. Miles also has over 15 years of work experience as a carpenter in commercial, residential, and modular construction. Mr. Miles is a freelance writer, specializing in educational writing. He is a contributing writer of *Agricultural Mechanics and Technology Systems*.

Reviewers

The author and publisher wish to thank the following industry and teaching professionals for their valuable input into the development of ***Math for Carpentry and Construction***.

Andrew Bell
Kotzebue Middle High School
Kotzebue, AK

Mark Enger
Renaissance High School
Meridian, ID

Jennifer Nichols
Hudson Valley Community College
Troy, NY

Bev Sroka
Milwaukee Area Technical College
Oak Creek, WI

James Wiater
Middlesex County Vocational School
East Brunswick, NJ

New to This Edition

The following changes have been made to the second edition of ***Math for Carpentry and Construction*** to strengthen the student's math skills so they can apply these skills in their trade.

- Section 6, *Material Estimating Activities*, will enhance the student's ability to calculate accurate materials quantities in order to supply competitive job estimates as well as order materials for the next phase of a project. Formulas and examples are provided for estimating materials needed for floor, wall, ceiling, roof, and stair frames. Activities are also provided for estimating concrete, roof finish, siding, insulation, and interior trim.
- Appendix A, *Construction Diagrams and Terms* consists of labeled diagrams illustrating foundation and floor frames, wall and roof frames, roof terminology, stair frames, and interior and exterior finish. A glossary of trade terms and components shown in these diagrams is also provided. This resource will be valuable for students to refer to while answering review questions in this text as well as on the job.
- *Understanding Measurement Tools Videos and Activities* consisting of 12 videos with worksheets and quizzes will help students learn and practice fundamental measurement skills they will use in class and on the job.

Features of the Textbook

The instructional design of this textbook includes student-focused learning tools to help you succeed. This visual guide highlights these features.

Section Opening Materials

Each section opener contains a table of contents of the units appearing in each section, along with a list of **Key Terms** to be learned in the section.

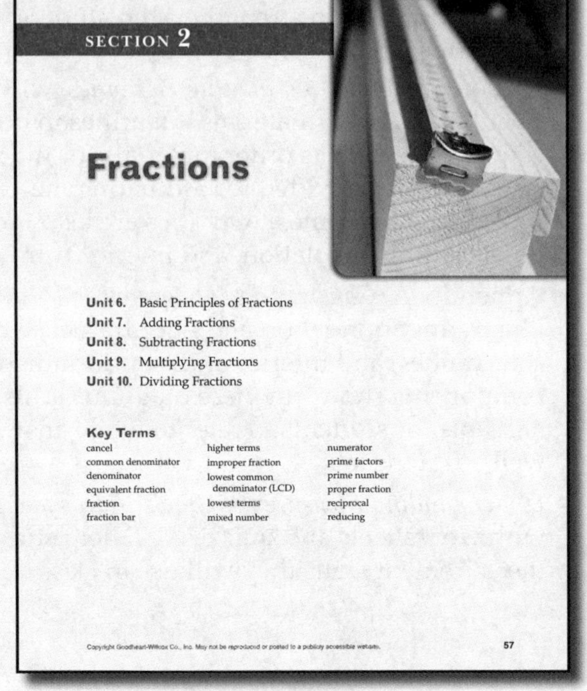

SECTION 2

Fractions

Unit 6. Basic Principles of Fractions
Unit 7. Adding Fractions
Unit 8. Subtracting Fractions
Unit 9. Multiplying Fractions
Unit 10. Dividing Fractions

Key Terms

cancel
common denominator
denominator
equivalent fraction
fraction
fraction bar
higher terms
improper fraction
lowest common denominator (LCD)
lowest terms
mixed number
numerator
prime factors
prime number
proper fraction
reciprocal
reducing

Copyright Goodheart-Willcox Co., Inc. May not be reproduced or posted to a publicly accessible website.

57

Unit Opening Material

Each unit opener lists **Objectives** that clearly identify the knowledge and skills to be gained when the unit is completed.

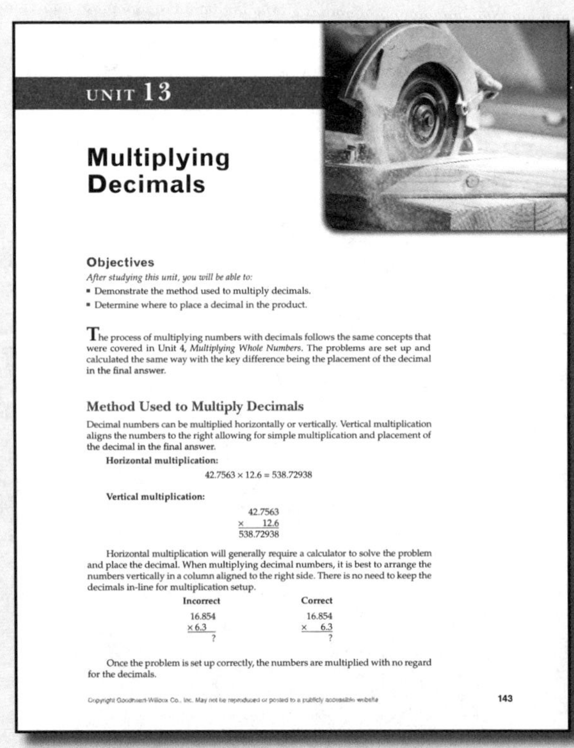

UNIT 13

Multiplying Decimals

Objectives

After studying this unit, you will be able to:

- Demonstrate the method used to multiply decimals.
- Determine where to place a decimal in the product.

The process of multiplying numbers with decimals follows the same concepts that were covered in Unit 4, *Multiplying Whole Numbers*. The problems are set up and calculated the same way with the key difference being the placement of the decimal in the final answer.

Method Used to Multiply Decimals

Decimal numbers can be multiplied horizontally or vertically. Vertical multiplication aligns the numbers to the right allowing for simple multiplication and placement of the decimal in the final answer.

Horizontal multiplication:

$$42.7563 \times 12.6 = 538.72938$$

Vertical multiplication:

$$\begin{array}{r} 42.7563 \\ \times \quad 12.6 \\ \hline 538.72938 \end{array}$$

Horizontal multiplication will generally require a calculator to solve the problem and place the decimal. When multiplying decimal numbers, it is best to arrange the numbers vertically in a column aligned to the right side. There is no need to keep the decimals in-line for multiplication setup.

Incorrect

$$\begin{array}{l} 16.854 \\ \times 6.3 \\ \hline \qquad ? \end{array}$$

Correct

$$\begin{array}{r} 16.854 \\ \times \quad 6.3 \\ \hline ? \end{array}$$

Once the problem is set up correctly, the numbers are multiplied with no regard for the decimals.

Copyright Goodheart-Willcox Co., Inc. May not be reproduced or posted to a publicly accessible website.

143

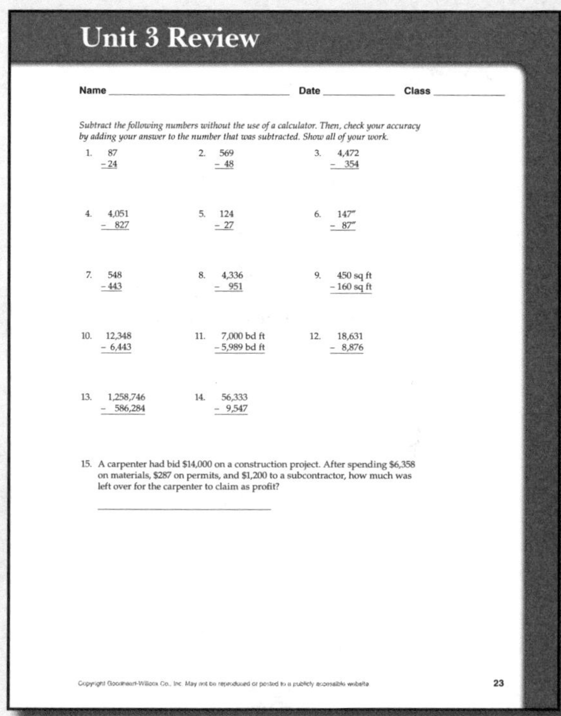

Unit 3 Review

Name ______________ Date ________ Class ________

Subtract the following numbers without the use of a calculator. Then, check your accuracy by adding your answer to the number that was subtracted. Show all of your work.

1. 87 − 24
2. 569 − 48
3. 4,472 − 354
4. 4,051 − 827
5. 124 − 27
6. 147″ − 87″
7. 548 − 443
8. 4,336 − 951
9. 450 sq ft − 160 sq ft
10. 12,348 − 6,443
11. 7,000 bd ft − 5,989 bd ft
12. 18,631 − 8,876
13. 1,258,746 − 586,284
14. 56,333 − 9,547
15. A carpenter had bid $14,000 on a construction project. After spending $6,358 on materials, $287 on permits, and $1,200 to a subcontractor, how much was left over for the carpenter to claim as profit?

____

23

Unit Reviews

End of unit material provides an opportunity for you to demonstrate knowledge and comprehension of unit material.

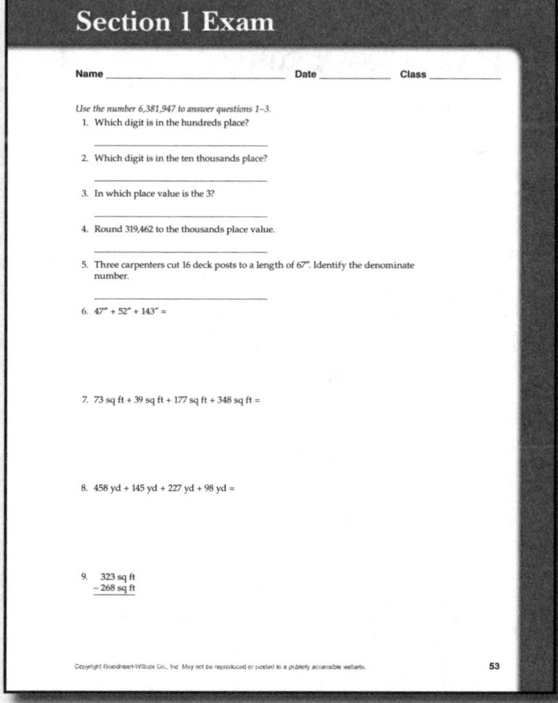

Section 1 Exam

Name ______________ Date ________ Class ________

Use the number 6,381,947 to answer questions 1–3.

1. Which digit is in the hundreds place?

   ____

2. Which digit is in the ten thousands place?

   ____

3. In which place value is the 3?

   ____

4. Round 319,462 to the thousands place value.

   ____

5. Three carpenters cut 16 deck posts to a length of 67″. Identify the denominate number.

   ____

6. 47″ + 52″ + 143″ =
7. 73 sq ft + 39 sq ft + 177 sq ft + 348 sq ft =
8. 458 yd + 145 yd + 227 yd + 98 yd =
9. 323 sq ft − 268 sq ft

53

Section Exams

End of section exams test retention of knowledge gained throughout units of each section.

Additional Features

Additional features are used throughout the body of each unit to further learning and knowledge. **Math Tips** underscore important points and provide additional easy-to-understand examples. **Examples** demonstrate the concept that has just been presented, showing all the work needed to solve a mathematical problem. **Carpentry Notes** help you explore industry situations and the math needed to solve problems on the job.

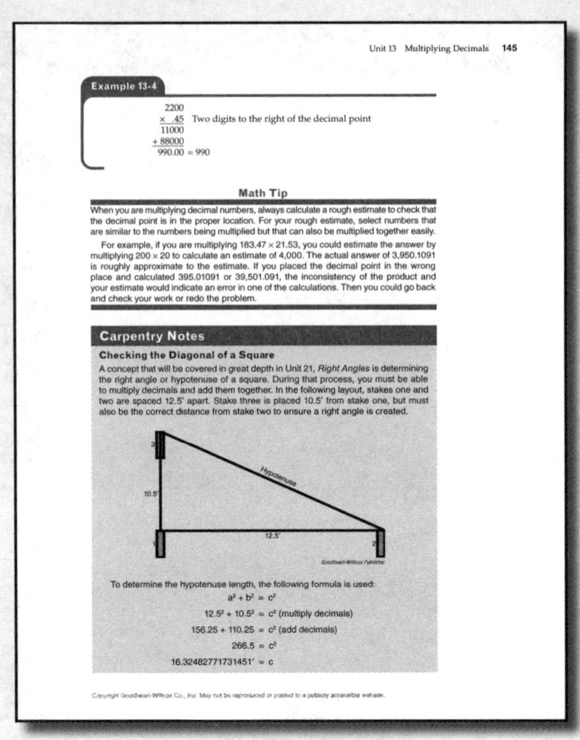

Unit 13 Multiplying Decimals 145

Example 13-4

```
  2200
×  .45   Two digits to the right of the decimal point
 11000
+88000
990.00 = 990
```

Math Tip

When you are multiplying decimal numbers, always calculate a rough estimate to check that the decimal point is in the proper location. For your rough estimate, select numbers that are similar to the numbers being multiplied but that can also be multiplied together easily.

For example, if you are multiplying 183.47×21.53, you could estimate the answer by multiplying 200×20 to calculate an estimate of 4,000. The actual answer of 3,950.1091 is roughly approximate to the estimate. If you placed the decimal point in the wrong place and calculated 395.01091 or 39,501.091, the inconsistency of the product and your estimate would indicate an error in one of the calculations. Then you could go back and check your work or redo the problem.

Carpentry Notes

Checking the Diagonal of a Square

A concept that will be covered in great depth in Unit 21, *Right Angles* is determining the right angle or hypotenuse of a square. During that process, you must be able to multiply decimals and add them together. In the following layout, stakes one and two are spaced 12.5′ apart. Stake three is placed 10.5′ from stake one, but must also be the correct distance from stake two to ensure a right angle is created.

To determine the hypotenuse length, the following formula is used:

$$a^2 + b^2 = c^2$$
$$12.5^2 + 10.5^2 = c^2 \text{ (multiply decimals)}$$
$$156.25 + 110.25 = c^2 \text{ (add decimals)}$$
$$266.5 = c^2$$
$$16.32482771731451' = c$$

Copyright Goodheart-Willcox Co., Inc. May not be reproduced or posted to a publicly accessible website.

An **Appendix** that provides diagrams as well as a glossary of trade terms and components has been added to this new edition. Construction terms are highlighted within the units, with a marginal note to refer to Appendix A as a reference.

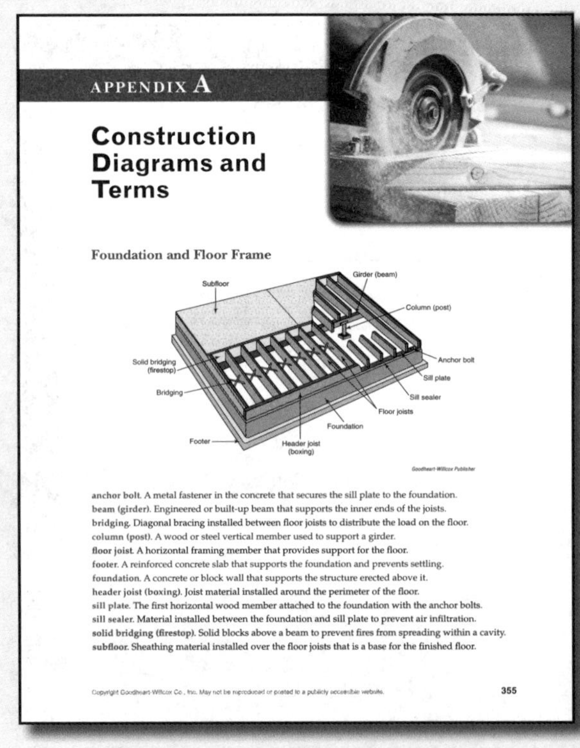

APPENDIX A

Construction Diagrams and Terms

Foundation and Floor Frame

anchor bolt. A metal fastener in the concrete that secures the sill plate to the foundation.
beam (girder). Engineered or built-up beam that supports the inner ends of the joists.
bridging. Diagonal bracing installed between floor joists to distribute the load on the floor.
column (post). A wood or steel vertical member used to support a girder.
floor joist. A horizontal framing member that provides support for the floor.
footer. A reinforced concrete slab that supports the foundation and prevents settling.
foundation. A concrete or block wall that supports the structure erected above it.
header joist (boxing). Joist material installed around the perimeter of the floor.
sill plate. The first horizontal wood member attached to the foundation with the anchor bolts.
sill sealer. Material installed between the foundation and sill plate to prevent air infiltration.
solid bridging (firestop). Solid blocks above a beam to prevent fires from spreading within a cavity.
subfloor. Sheathing material installed over the floor joists that is a base for the finished floor.

Copyright Goodheart-Willcox Co., Inc. May not be reproduced or posted to a publicly accessible website. 355

Expanding Your Learning

Activities have been added to apply math skills in preparing estimates for material quantities needed for a project.

ACTIVITY 1

Estimating Concrete for Slabs, Footers, and Walls

Objective

After studying this section, you will be able to:

- Estimate concrete for slabs, footers, and walls.

Concrete is ordered for slabs (sidewalks, patios, driveways, cellar floors), footers below a foundation wall, or vertical walls. Concrete is calculated by its volume taking into account the projects thickness (t), width (w), and length (l) in feet (see Unit 18, *Volume Measurement*). The formula used to calculate concrete quantities is:

$$\frac{t' \times w' \times l'}{27} = \text{cubic yards of concrete}$$

When the thickness, width, and length are calculated in feet, the answer will be in cubic feet. Since concrete is ordered and sold by the cubic yard, it is necessary to convert the answer to cubic yards. There are 27 cubic feet within a cubic yard.

Example 22-1

Calculate the amount of concrete needed for a project measuring 6″ thick × 14′ wide × 22′ long.

$$\frac{.5' \times 14' \times 22'}{27} = \frac{154 \text{ cu ft}}{27} = 5.7 \text{ cubic yards of concrete}$$

Math Tip

When the measurements for concrete are inserted into the formula, any variable that is not an even foot must be expressed as a decimal foot (see Unit 15, *Linear Measurement*). For example, 4″ = .333′.

Slabs

Once the measurements of a slab are determined, they are inserted into the formula and the calculations are made to determine the cubic yards needed. A good practice to follow is to add 5% onto your total to allow for possible spillage or over-excavation of the site.

283

Name ____________ Date ________ Class ________

ACTIVITY 1

Estimating Concrete for Slabs, Footers, and Walls

Solve the following problems related to estimating concrete. Show all of your work and round answers to the nearest tenth.

1. Estimate concrete for a patio measuring 4″ thick × 12′ wide × 18′ long.
   ____
2. Estimate concrete for a driveway measuring 6″ thick × 22′ wide × 23′ 6″ long.
   ____
3. Estimate concrete for a sidewalk measuring 4″ thick × 3′ wide × 44′ long.
   ____

287

TOOLS FOR STUDENT AND INSTRUCTOR SUCCESS

Student Tools

Student Text

Math for Carpentry and Construction is a write-in textbook that provides a wealth of examples and exercises for an in-depth learning experience. This edition includes a new section with estimating activities as well as an industry-related glossary in the Appendix.

G-W Digital Companion

For digital users, e-flash cards and vocabulary exercises allow interaction with content to create opportunities to increase achievement.

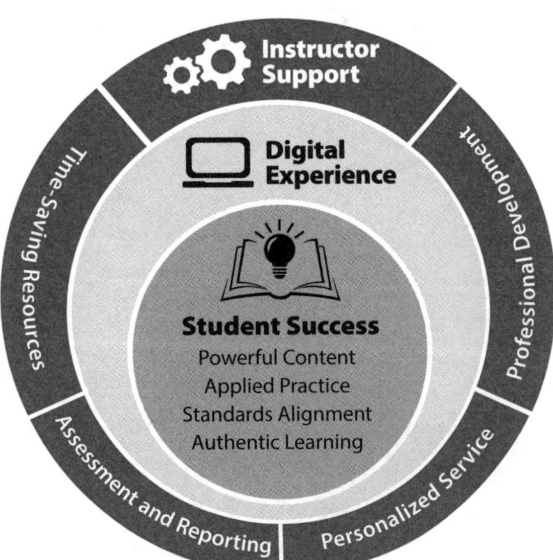

Instructor Tools

LMS Integration

Integrate Goodheart-Willcox content within your Learning Management System for a seamless user experience for both you and your students. EduHub LMS–ready content in Common Cartridge® format facilitates single sign-on integration and gives you control of student enrollment and data. With a Common Cartridge integration, you can access the LMS features and tools you are accustomed to using and G-W course resources in one convenient location—your LMS.

G-W Common Cartridge provides a complete learning package for you and your students. The included digital resources help your students remain engaged and learn effectively:

- **eBook**
- **Drill and Practice** vocabulary activities

When you incorporate G-W content into your courses via Common Cartridge, you have the flexibility to customize and structure the content to meet the educational needs of your students. You may also choose to add your own content to the course.

For instructors, the Common Cartridge includes the Online Instructor Resources. QTI® question banks are available within the Online Instructor Resources for import into your LMS. These prebuilt assessments help you measure student knowledge and track results in your LMS gradebook. Questions and tests can be customized to meet your assessment needs.

Online Instructor Resources (OIR)

- The **Instructor Resources** provide instructors with time-saving preparation tools such as answer keys, editable lesson plans, and other teaching aids.
- **Instructor's Presentations for PowerPoint®** are fully customizable, richly illustrated slides that help you teach and visually reinforce the key concepts from each unit.
- Administer and manage assessments to meet your classroom needs using **Assessment Software with Question Banks**, which include hundreds of matching, completion, multiple choice, and short answer questions to assess student knowledge of the content in each unit.

See **www.g-w.com/math-for-carpentry-construction-2024** for a list of all available resources.

Professional Development

- Expert content specialists
- Research-based pedagogy and instructional practices
- Options for virtual and in-person Professional Development

Brief Contents

Section 1 Whole Numbers

Section 2 Fractions

Section 3 Decimals

Section 4 Measurement

Section 5 Percentages and Right Angles

Section 6 Material Estimating Activities

Contents

SECTION 1
Whole Numbers

SECTION 2
Fractions

SECTION 6
Material Estimating Activities

Feature Contents

Carpentry Notes

SECTION 1

Whole Numbers

Unit 1. Basic Principles of Whole Numbers
Unit 2. Adding Whole Numbers
Unit 3. Subtracting Whole Numbers
Unit 4. Multiplying Whole Numbers
Unit 5. Dividing Whole Numbers

Key Terms

addition
borrowing
carrying
decimal number system
denominate number
difference
digits
dividend
division
division sign (÷)
divisor
equals sign (=)
minus sign (–)
multiplication
multiplication sign (×)
place value
plus sign (+)
product
quotient
remainder
rounding
square
subtraction
sum
whole number

UNIT 1

Basic Principles of Whole Numbers

Objectives

After studying this unit, you will be able to:

- Identify the place value of whole numbers.
- Round whole numbers to a given place value.
- Recognize denominate numbers.

Carpenters and other workers in the building construction trades must develop a mastery of mathematics principles, operations, and formulas. Every aspect of the construction industry requires an understanding and accurate use of mathematical concepts. As a carpenter, you will need to be skilled in tasks such as reading a tape measure, estimating materials, and calculating rafter line length, just to name a few. Calculators can be an efficient tool when performing math operations, but they are not always available, or practical, on a construction site. Because of this, it is essential that carpenters learn basic math principles and develop the ability to solve operations without the use of a calculator.

Place Values

The **decimal number system**, also known as the Arabic number system, uses 10 as its base unit. This system consists of 10 different digits: 0, 1, 2, 3, 4, 5, 6, 7, 8, and 9. **Digits** are symbols that are used to create everyday numbers.

Numerical values are expressed by placing a combination of digits together to create a whole number. A **whole number** is a numeral that stands alone as a unit and can contain multiple digits. Whole numbers do not include a fraction or a decimal. In the decimal number system, the value of the digit is determined by its location in the whole number. This value is called the **place value**. The following chart shows the place value of each digit within the number 5,694.

Goodheart-Willcox Publisher

Add the values of each digit to determine the value of the whole number.

The digit 5 is in the thousands column	$5 \times 1{,}000$ =	5,000
The digit 6 is in the hundreds column	6×100 =	600
The digit 9 is in the tens column	9×10 =	90
The digit 4 is in the ones column	4×1 =	+ 4
	=	5,694

Math Tip

To make numbers easier to read, a comma is inserted after every third digit counting from the right.

Example 1-1

What are the place values of the digits in the number 5,298,347?

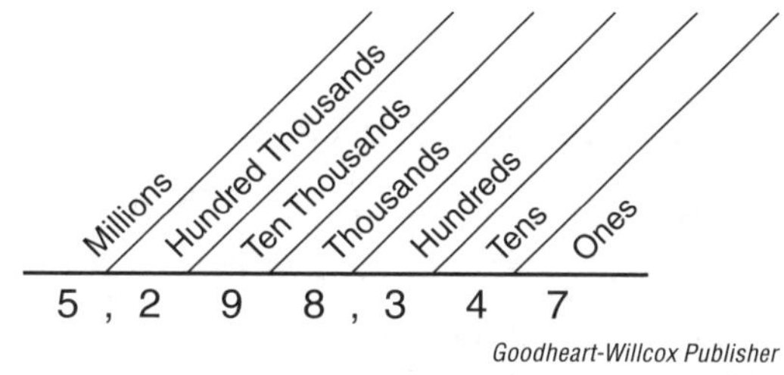

Goodheart-Willcox Publisher

Example 1-2

Sometimes a zero is one of the digits within a whole number. Regardless of where it is located in the number, its place value remains zero.

What is the place value of the zero in the whole number 2,084?

The zero is in the hundreds column, so then $0 \times 100 = 0$.

The whole number would be read as two thousand eighty-four.

Rounding Whole Numbers

There are times when whole numbers do not need to be exact. When calculating material estimates for a project or cost estimates for a quote, numbers can be rounded to a specific place value. **Rounding** numbers is the process of increasing or decreasing the value of the number to a specific place value. When calculating the area of the roof in the following image, the 1,382 sq ft roof would be rounded to 1,400 sq ft for estimating or purchasing the material. Because shingles are sold by the square (100 sq ft), the area of the roof is rounded up to the next full square.

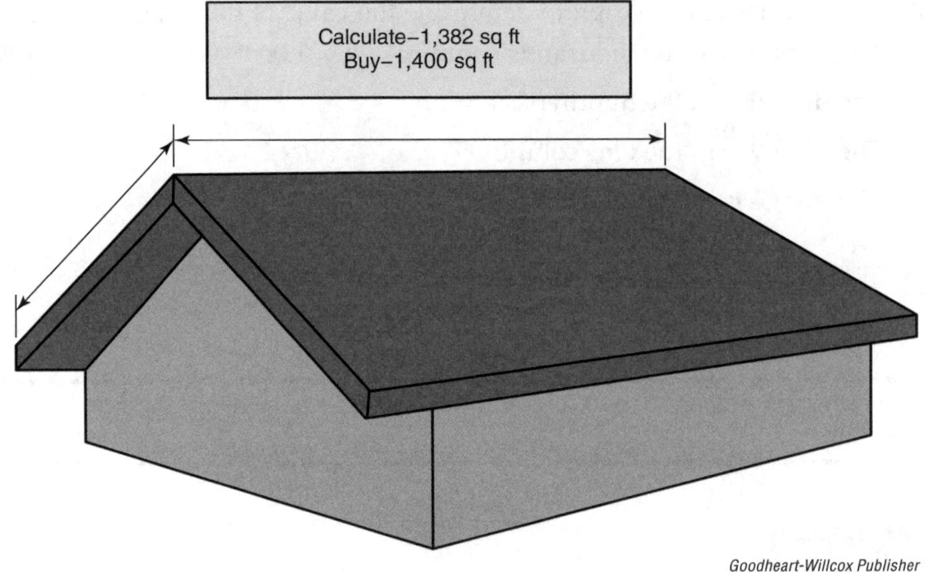

Goodheart-Willcox Publisher

When calculating material estimates for a project, numbers are generally rounded up to account for how materials are sold or for waste factor. Other times, numbers can be rounded up or down to a specific place value depending on the situation.

Steps for Rounding to a Nearest Place Value

1. Identify the place value to which the number is being rounded.
2. Identify the digit to the right of that place value.
3. If the digit to the right is 5 or greater, increase the place value number up by one (round up). If the digit to the right is 4 or less, do not change the place value number.
4. Change the remaining digits to the right to zeros.

Example 1-3

Round the number 34,673 to the thousands place value.

The number in the thousands place value is 4.

3**4**,673

The digit to the right of the 4 is 6.

34,**6**73

Since 6 is greater than 5, the 4 is rounded up by one, making it 5. The remaining numbers to the right are changed to zero.

3**5,000**

Example 1-4

Round the number 16,841 to the hundreds place value.

The number in the hundreds place value is 8.

16,**8**41

The digit to the right of the 8 is 4.

16,8**4**1

Since the 4 is less than 5, the 8 is not changed but the remaining numbers to the right are changed to zero.

16,8**00**

Denominate Numbers

When working in construction, numbers will often be associated with a unit of measurement. Numbers consisting of a numeric value and a unit of measure are referred to as **denominate numbers**. Carpenters work with denominate numbers in every phase of the construction process from site prep to framing to final finish. For example, 32′ and 64′ are denominate numbers because they include a value (32 and 64) and a unit of measure (feet). Other examples include:

- 15 lb
- 64 sq ft
- 3 hours, 14 minutes, 45 seconds
- 2 tons, 600 pounds

Carpentry Notes

Paint Estimation

During the finishing stages of construction, carpenters take room measurements and purchase the appropriate quantity of paint for the walls and ceilings. This is one occasion when carpenters take exact calculations, and then round the number up. Rounding up will help cover any extra paint needed for touch-ups or mistakes. If 1 gallon of paint will cover an area of 400 sq ft, how many gallons of paint are needed for a 764 sq ft room?

To round up 764 sq ft to the nearest hundreds, look at the number to the right of the hundreds place value. As 6 is greater than 5, then round up the hundreds place value to 8, which gives you 800 sq ft. Since 1 gallon of paint will cover an area of 400 sq ft, 2 gallons of paint are needed for this project.

Work Space/Notes

Unit 1 Review

Name ______________________ **Date** __________ **Class** __________

Use the number 4,784,523 to answer the following questions.

1. Which digit is in the hundreds place?

____

2. Which digit is in the ten thousands place?

____

3. Which digit is in the tens place?

____

4. Which digit is in the hundred thousands place?

____

5. Which digit is in the thousands place?

____

Use the number 4,721,593 to answer the following questions.

6. In which place value is the 3?

____

7. In which place value is the 7?

____

8. In which place value is the 4?

____

9. In which place value is the 9?

____

10. In which place value is the 1?

____

Round the following whole numbers to the indicated place value.

11. Round 172,839 to the ten thousands place value.

____

12. Round 74,613 to the thousands place value.

____

13. Round 93,827 to the tens place value.

____

14. Round 17,464,594 to the millions place value.

____

15. Round 8,462 to the hundreds place value.

____

16. Round 729,381 to the hundred thousands place value.

____

17. Round 943 to the hundreds place value.

____

18. Round 91,735 to the tens place value.

____

19. Round 31,723,468 to the ten thousands place value.

____

20. Round 6,527,914 to the thousands place value.

____

Identify the denominate numbers in the following problems.

*See Appendix A
Wall Frame

21. A carpenter cuts 64 boards to a length of 93″ for wall ***studs****.

____

*See Appendix A
Foundation and Floor Frame

22. The floor plan called for 8 ***floor joists**** to be installed 16″ on center.

____

23. The dimensions on the floor plan are 16′ by 14′ for the 2 bedrooms.

____

24. The estimated time for 4 workers to shingle the roof was 6 hours, 30 minutes.

____

25. The 7-acre property building site will have 3 buildings on the north end.

____

UNIT 2

Adding Whole Numbers

Objectives

After studying this unit, you will be able to:

- Practice methods used to add whole numbers.
- Demonstrate carrying in the addition process.
- Solve addition problems with denominate numbers.

The four basic operations of arithmetic are addition, subtraction, multiplication, and division. Addition will be discussed in this unit. **Addition** is the process of combining two or more numbers to obtain a combined quantity called the **sum**. A **plus sign (+)** is used to indicate the numbers being added.

Method Used to Add Whole Numbers

When adding denominate numbers, the unit of measure is applied to the answers as well. As shown in the following image, the sum of adding four bags of cement and another two bags adds up to six bags of cement.

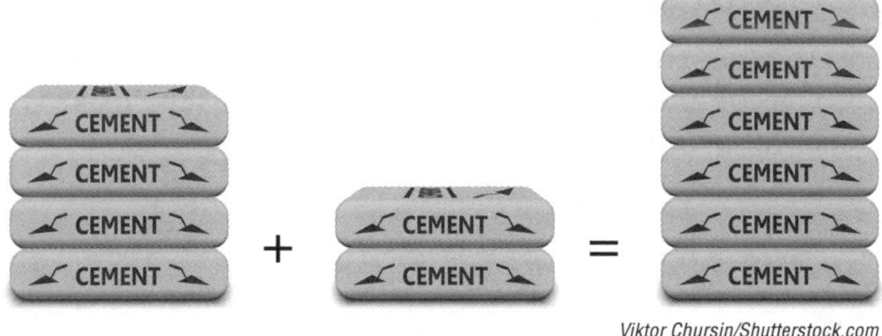

Viktor Chursin/Shutterstock.com

Numbers can be added either horizontally or vertically.

Horizontal addition:

$$4 + 2 = 6$$

Vertical addition:

$$\begin{array}{r} 4 \\ +\,2 \\ \hline 6 \end{array}$$

Math Tip

When performing math operations vertically, the **equals sign (=)** is replaced with a horizontal line located below the numbers being added.

Horizontal adding is useful when adding simple, single-digit numbers, but when numbers become two or more digits long, vertical adding is the more practical method to use. Regardless of the size of each number, the vertical addition problem is always set up with the numbers aligned on their right side. Equivalent place values of each number must line up in the same column and are then added together vertically.

Example 2-1

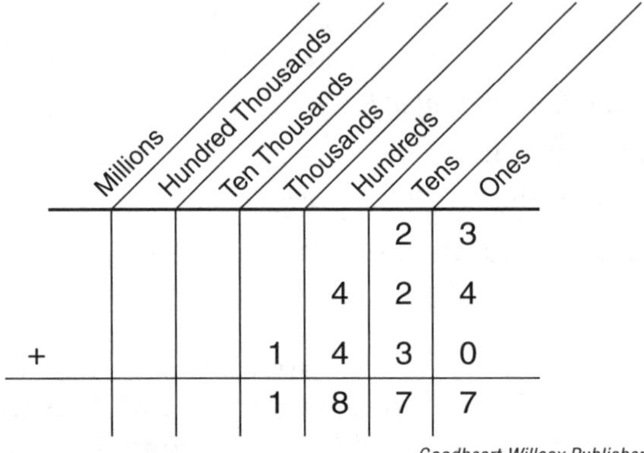

Goodheart-Willcox Publisher

The numbers in the ones column are added together (3 + 4 + 0 = 7), the tens column (2 + 2 + 3 = 7), the hundreds column (4 + 4 = 8), and the thousands column (1 + 0 = 1). The sum of each column is a single digit, making the final answer 1,877.

Carrying in Addition

When adding individual columns, the answer must always remain a single digit. If the sum of adding a column is 10 or greater, **carrying** must be used to keep a single digit in each place value of the sum.

Example 2-2

$$\begin{array}{r} {\scriptstyle 1} \\ 47 \\ +\ 128 \\ \hline 175 \end{array}$$

The sum of adding (7 + 8) in the ones column produces a double-digit answer (15). The 5 remains in the ones column and the 10 is carried over to the next column. Even though 10 is being carried over, it is written as 1 since the next column place value is ten times greater than the ones column.

After the ones column is added and 10 has been carried over, the tens column is added (1 + 4 + 2 = 7). Then, the hundreds column is added (1 + 0 = 1). The final answer is 175.

Example 2-3

As whole numbers become larger, carrying is used as needed when each column is added.

$$\begin{array}{r} {\scriptstyle 2\ 1 2} \\ 5{,}864 \\ 629 \\ +\ 578 \\ \hline 7{,}071 \end{array}$$

Ones column:	4 + 9 + 8 = 21	The 1 remains and the 20 is carried over
Tens column:	2 + 6 + 2 + 7 = 17	The 7 remains and the 10 is carried over
Hundreds column:	1 + 8 + 6 + 5 = 20	The 0 remains and the 20 is carried over
Thousands column:	2 + 5 = 7	The final answer is 7,071

Adding Denominate Numbers

As you learned in Unit 1, *Basic Principles of Whole Numbers*, a denominate number is a whole number with a unit of measure associated with it (for example, 18 inches). Denominate numbers follow the same process used for adding whole numbers, but additional rules must be applied.

Rule #1: Denominate numbers must have the same type of unit measure associated with them.

$$\begin{array}{r} 53 \text{ in} \\ +\ 27 \text{ lb} \\ \hline ? \end{array}$$

These numbers cannot be added because the units of measure are not the same. Inches cannot be added to pounds.

$$\begin{array}{r} 6'' \\ +\ 8'' \\ \hline 14'' \end{array}$$

When denominate numbers have the same unit of measurement (inches in the preceding example), they can be added together. The sum will have the same unit of measurement applied to it.

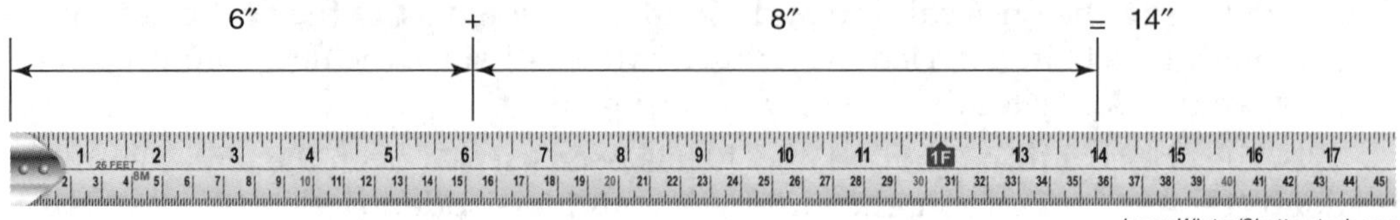

Jason Winter/Shutterstock.com

Rule #2: Units of measure in compound denominate numbers must line up with like units.

$$\begin{array}{lrrr} & 4\text{ yd} & 1\text{ ft} & 3\text{ in} \\ + & & 1\text{ ft} & 2\text{ in} \\ \hline & 4\text{ yd} & 2\text{ ft} & 5\text{ in} \end{array}$$

Carpentry Notes

Mental Math

Often, carpenters on a jobsite are required to add multiple numbers without the aid of a calculator. A good practice to follow is to group together numbers that will add up to 10, or units of 10, and then determine the sum from there.

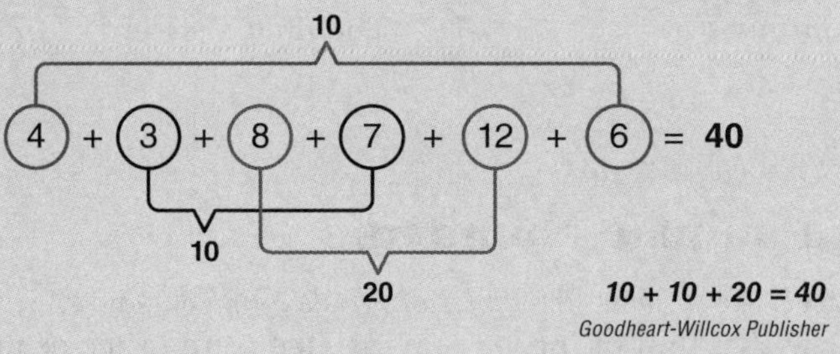

Goodheart-Willcox Publisher

Unit 2 Review

Name ______________________ **Date** __________ **Class** __________

Add the following numbers without the use of a calculator. Show all of your work.

1. 73 + 24

2. 51 + 27

3. 96 + 47

4. 89 + 25

5. 74 + 538

6. 34 + 299

7. 657 + 1,915

8. 581 + 2,175

9. 2,348 + 7,999

10. 5,286 + 6,123

11. 27″ 49″ + 124″

12. 145 sq ft 27 sq ft 182 sq ft 210 sq ft + 445 sq ft

13. 1,528 yd 23 yd 462 yd + 17 yd

14. 23 sq ft 156 sq ft 1,688 sq ft + 14,295 sq ft

15. $4,625 $2,200 $580 + $1,225

16. A carpenter must purchase ***sill seal**** for a 28′ × 64′ foundation. How many lineal feet of material is needed to seal the perimeter of the foundation?

____

***See Appendix A**
Foundation and Floor Frame

17. Finish trim needs to be ordered for a new home under construction. Using the following chart, determine the total lineal feet required for each of the trim components.

Room	Base	Casing	Crown	Chair Rail
Living Room	68′	128′	84′	N/A
Dining Room	48′	66′	56′	48′
Kitchen	18′	97′	N/A	N/A
Den	46′	93′	52′	46′
Bedroom 1	52′	108′	64′	N/A
Bedroom 2	46′	92′	N/A	N/A
Bedroom 3	38′	77′	N/A	N/A
Bath 1	18′	94′	N/A	N/A
Bath 2	8′	82′	N/A	N/A
Total Lineal Feet				

Goodheart-Willcox Publisher

18. A general contractor hired a painting crew to paint a newly constructed home. Three painters worked 8 hours a day for five days, two painters worked 8 hours a day for three days, and one painter worked 5 hours a day for five days. How many total hours did the painting crew work on the job?

____

19. A carpenter purchasing materials for a project paid $1,286; $3,628; $2,340; and $3,688 to four different vendors. What was the total cost of the materials?

____

Name ______________________ Date ____________ Class ____________

20. A new home construction project is estimated to require 5 cu yd of concrete for the footer, 30 cu yd for the foundation walls, 16 cu yd for the basement floor, 8 cu yd for the driveway, and 5 cu yd for the sidewalks. How many cubic yards of concrete will be used for the entire project?

____

21. A contractor is building three homes in a development and needs to order shingles for delivery. The three roofs each require 1,240 sq ft, 1,860 sq ft, and 1,440 sq ft of architectural shingles. How many square feet of shingles must the contractor order?

____

22. Using the dimensions shown in the partial foundation plan, what is the distance from the edge of the foundation wall on the left to that on the right?

____

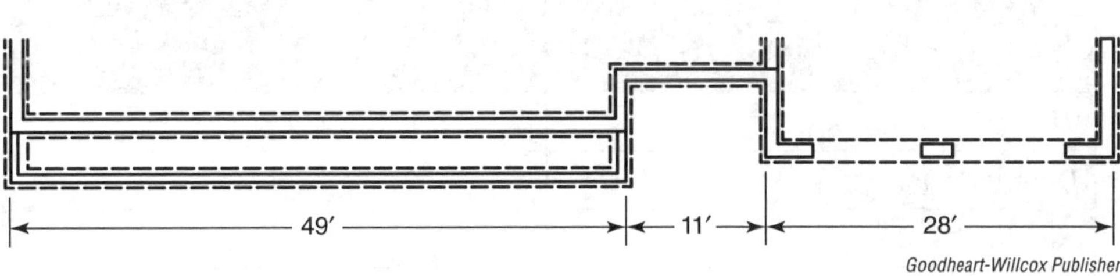

Goodheart-Willcox Publisher

*See Appendix A
Wall Frame

23. Material needs to be ordered to build ***headers**** for an addition. How many inches of boards are needed to build two 36″ headers, five 48″ headers, and one 72″ header? Note that all headers are built 2 boards wide.

____

24. The following rooms will have hardwood flooring installed in them. How many square feet of flooring need to be ordered?

Living room—258 sq ft

Dining room—180 sq ft

Den—182 sq ft

Foyer—112 sq ft

____

25. An electrical contractor must order electrical boxes for the following rooms. Determine the total number of each type of box needed.

Room	One-Gang Square	Two-Gang Square	Round Ceiling
Living Room	9	2	2
Dining Room	7	2	1
Kitchen	11	4	2
Bedroom 1	9	2	1
Bedroom 2	8	1	1
Bath 1	3	2	1
Bath 2	2	2	1
Total			

Goodheart-Willcox Publisher

Name ______________________ Date ____________ Class ____________

26. Stock kitchen cabinets were ordered to be installed from corner to corner along the north wall of a kitchen. Two 36″ cabinets, two 24″ cabinets, one 30″ cabinet, and one 18″ cabinet are to be installed. What is the length of the wall where they will be installed?

____

27. In the following image, how many lineal feet of pressure-treated boards are needed for the ***sill plate**** on the foundation?

*See Appendix A
Foundation and Floor Frame

____

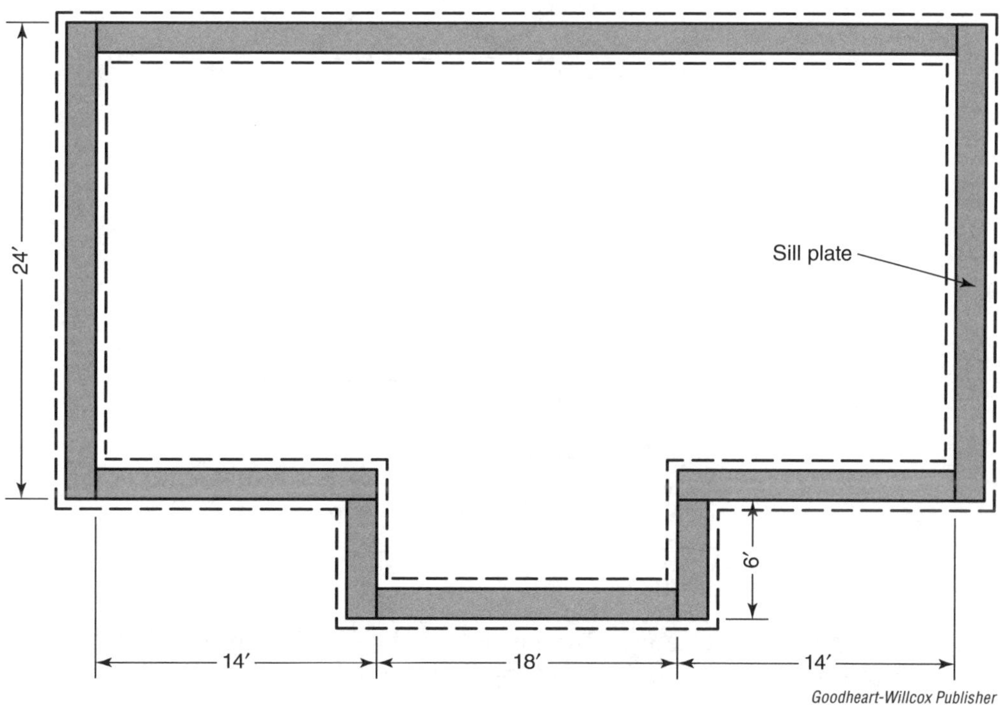

28. A cabinetmaker needs to order maple to build three future projects. The board foot (bd ft) requirements for the three projects are 48 bd ft for a corner cabinet, 56 bd ft for a hope chest, and 112 bd ft for an armoire. How many board feet of maple does the cabinetmaker need to order?

____

*See Appendix A

Foundation and Floor Frame

29. In the following image, what is the distance from the bottom of the ***footer**** to the floor line?

____

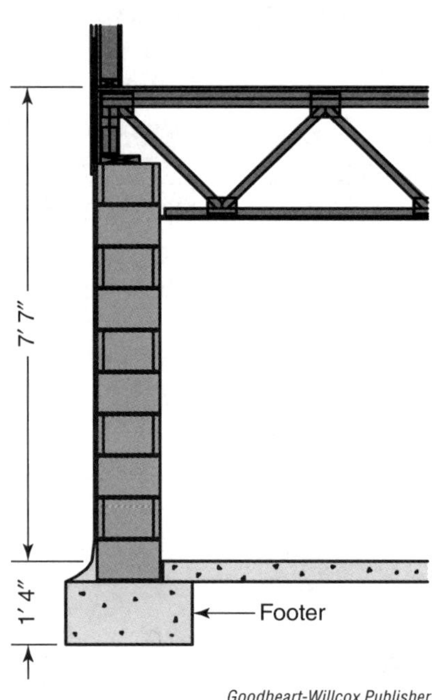

Goodheart-Willcox Publisher

30. A building lot has perimeter dimensions of 110′, 116′, 174′, and 168′. What is the perimeter or total distance around the building lot?

____

UNIT 3

Subtracting Whole Numbers

Objectives

After studying this unit, you will be able to:

- Practice methods used to subtract whole numbers.
- Demonstrate the process of borrowing in subtraction.
- Check subtraction by adding.
- Solve subtraction problems with denominate numbers.

The second of the four basic mathematical operations is subtraction. Unlike addition, which combines units, **subtraction** is the process of removing units to obtain a smaller quantity. This smaller quantity is referred to as the **difference**. A **minus sign (–)** is used to indicate the numbers being subtracted. For example, if a carpenter uses 4 boards from a pile of 12 boards, the difference would be 8 boards.

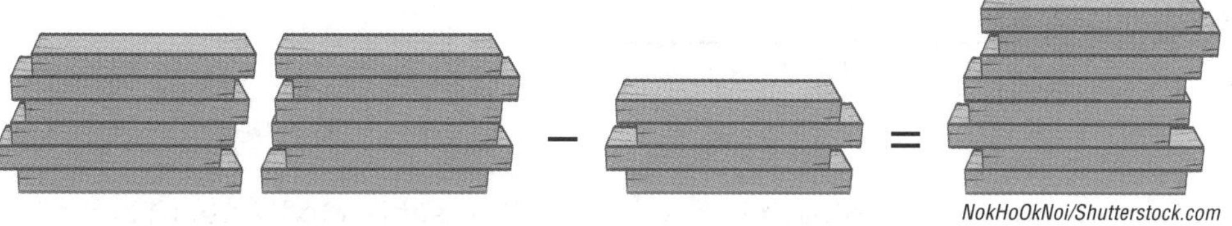

NokHoOkNoi/Shutterstock.com

Method Used to Subtract Whole Numbers

When subtracting numbers, they can be set up and subtracted either horizontally or vertically. Using the boards example:

Horizontal subtraction:

$$12 - 4 = 8$$

Vertical subtraction:

$$\begin{array}{r} 12 \\ -\ 4 \\ \hline 8 \end{array}$$

As with addition, horizontal subtraction may be used on single-digit numbers, but vertical subtraction is the preferred method with larger numbers. Problems are always set up with the numbers aligned on the right side, with the larger value as the top number in the operation. Equivalent place values of each number must line up in the same column and are then subtracted vertically working right to left.

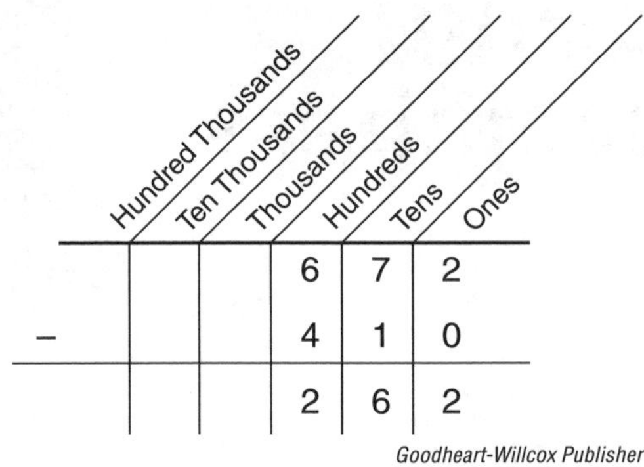

Goodheart-Willcox Publisher

The numbers in the ones column are subtracted (2 – 0 = 2), the tens column (7 – 1 = 6), and the hundreds column (6 – 4 = 2). The difference of subtracting each column was a single digit making the final answer 262.

Borrowing in Subtraction

When subtracting across the place value columns, in some instances the top number may be less than the lower number. When this occurs, borrowing must be used. **Borrowing** is the process of taking one unit from the place value to the left of the column being subtracted.

Example 3-1

$$\begin{array}{r} 65 \\ -\,27 \\ \hline \end{array}$$

In the ones column, 7 is larger than 5, so it cannot be subtracted without borrowing. In this example, one must be borrowed from the tens column and added to the 5. Since the value of the 6 in the tens column is actually 60, when one unit is moved to the ones column, 10 will be added to the 5, making its value 15.

$$\begin{array}{r} {\scriptstyle 5} \\ \not{6}{}^{1}5 \\ -\,27 \\ \hline 38 \end{array}$$

The problem can now be completed by subtracting the ones column (15 – 7 = 8) and the tens column (5 – 2 = 3), making the difference 38.

Math Tip

Borrowing may need to be done multiple times when solving a subtraction problem. When borrowing, always make sure it is done working right to left.

Example 3-2

In this example, the numbers being subtracted in the ones and tens columns are larger than the numbers they are being subtracted from.

$$\begin{array}{r} 301 \\ -\,146 \\ \hline \end{array}$$

In the ones column, the 6 cannot be subtracted from 1. Borrowing from the tens column is not possible because it has a zero value. When this occurs, move left again until a column is found with a value that can be borrowed from. In this case, one unit can be borrowed from the hundreds column. The 3 in the hundreds column is reduced to 2, raising the 0 in the tens column to 10.

$$\begin{array}{r} \overset{2}{\cancel{3}}\overset{1}{0}1 \\ -\,146 \\ \hline \end{array}$$

Now a unit can be borrowed from the tens column, raising the ones column to 11. With all the borrowing complete, the operation can be done to find the difference.

$$\begin{array}{r} \overset{2}{\cancel{3}}\overset{9}{\cancel{10}}\overset{1}{1} \\ -\,146 \\ \hline 155 \end{array}$$

Ones column:	11 – 6 = 5
Tens column:	9 – 4 = 5
Hundreds column:	2 – 1 = 1

Checking Subtraction by Adding

An answer to a subtraction problem can be checked for accuracy by reversing the operation and adding the answer to the number that was subtracted. The sum should produce the original top number.

Using this method to check the problem in Example 2 would look like this:

$$\begin{array}{r} 301 \\ -\,146 \\ \hline 155 \end{array} \qquad \begin{array}{r} 155 \\ +\,146 \\ \hline 301 \end{array}$$

Subtracting Denominate Numbers

Subtracting denominate numbers follows the same rules covered in Unit 2, *Adding Whole Numbers,* when adding denominate numbers.

1. To be subtracted, denominate numbers must have the same type of unit.
2. Compound denominate numbers must line up with like units.
3. Once denominate numbers are subtracted, their units of measure may need to be converted to simplify them.

$$\begin{array}{rrr} 6\text{ yd} & 2\text{ ft} & 10\text{ in} \\ -\,2\text{ yd} & 1\text{ ft} & 4\text{ in} \\ \hline 4\text{ yd} & 1\text{ ft} & 6\text{ in} \end{array}$$

Work Space/Notes

Unit 3 Review

Name ______________________ Date ____________ Class ____________

Subtract the following numbers without the use of a calculator. Then, check your accuracy by adding your answer to the number that was subtracted. Show all of your work.

1. 87 − 24
2. 569 − 48
3. 4,472 − 354
4. 4,051 − 827
5. 124 − 27
6. 147″ − 87″
7. 548 − 443
8. 4,336 − 951
9. 450 sq ft − 160 sq ft
10. 12,348 − 6,443
11. 7,000 bd ft − 5,989 bd ft
12. 18,631 − 8,876
13. 1,258,746 − 586,284
14. 56,333 − 9,547

15. A carpenter had bid $14,000 on a construction project. After spending $6,358 on materials, $287 on permits, and $1,200 to a subcontractor, how much was left over for the carpenter to claim as profit?

____

16. If two 32″ cuts were made from a 144″ board, how many inches would be left over?

____

*See Appendix A
Foundation and Floor Frames

17. A floor system requires 62 sheets of tongue-and-groove OSB for the ***subfloor****, and there are already 38 sheets on-site. How many additional sheets need to be ordered?

____

Use the following figure to find the missing distance. Each question has a different set of measurements. Window and door placement is generally measured on a set of floor plans from the outside corners to the center of the unit. Partition walls can also be measured to their center.

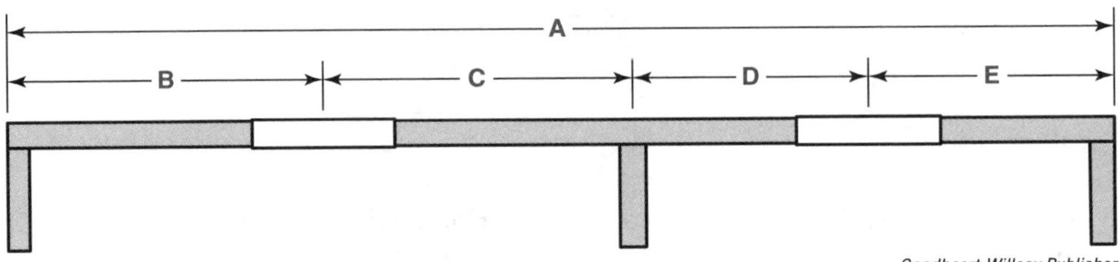

Goodheart-Willcox Publisher

	Distance A	Distance B	Distance C	Distance D	Distance E
18.	24′	7′	7′	5′	
19.	32′	9′	10′		6′
20.	26′	8′		5′	4′
21.	28′		9′	5′	5′

Goodheart-Willcox Publisher

Name ______________________ Date ____________ Class ____________

22. A building supply company had 380 sheets of drywall in stock. In one week, that stock was reduced by 90 sheets to a commercial contractor, 27 sheets to a local residential contractor, and 14 sheets that were returned to the manufacturer because of damage. At the end of the week, how many sheets were still in stock?

____

23. A local contractor bid a job to take 95 total hours. Monday, 4 of his workers worked an 8-hour day. Tuesday, 3 worked an 8-hour day and one worked a 6-hour day. Wednesday, 2 worked a 7-hour day and one worked 4 hours. How many hours does the contractor have remaining to finish the job?

____

24. An excavator must dig a trench 6′9″ below an established benchmark. Currently, the excavator has dug 4′3″ below the benchmark. Referencing the following figure, how much deeper must the excavator dig?

____

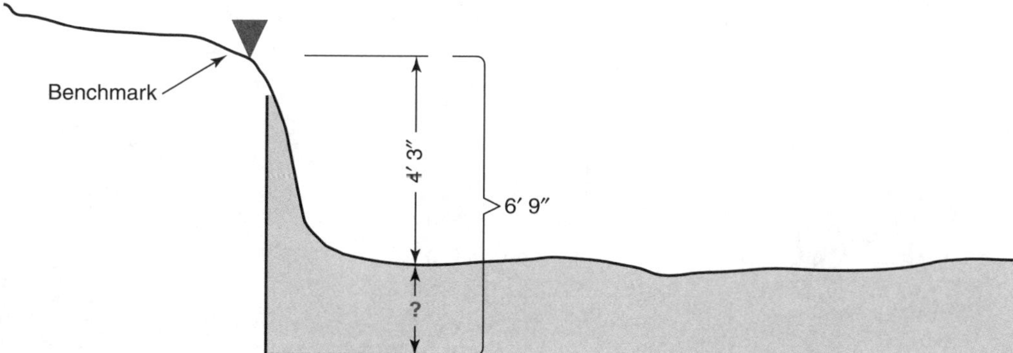

Goodheart-Willcox Publisher

25. A flooring contractor purchased 7,000 sq ft of hardwood floor for three homes. The first house will require 2,200 sq ft, and the other two will use 1,650 sq ft and 1,475 sq ft, respectively. How much hardwood flooring will remain after these three jobs are completed?

____

UNIT 4

Multiplying Whole Numbers

Objectives

After studying this unit, you will be able to:

- Practice the method used to multiply whole numbers.
- Demonstrate carrying in multiplication problems.
- Check multiplication problems by reversing the position of the numbers in the problem.
- Solve multiplication problems with denominate numbers.

Multiplication is the third of the four basic mathematical operations. **Multiplication** can be thought of as repeat addition, or a quick method of adding the same number together multiple times. For example, if a contractor was building five homes from the same set of prints, and each home required four smoke detectors, the contractor would need 20 detectors. This can be determined through addition.

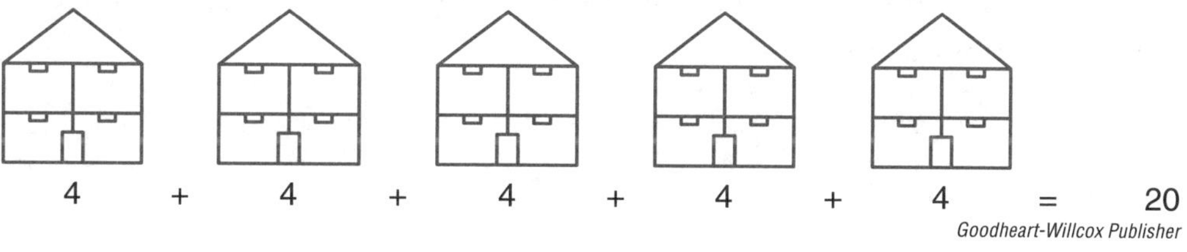

Goodheart-Willcox Publisher

A quicker way to solve this problem is through multiplication. This problem can then be viewed as 4 taken 5 times, or 5 times the number 4. Multiplication problems are identified by the **multiplication sign** (×). The answer to a problem is called the **product**.

$$4 \times 5 = 20$$

The easiest way to perform simple multiplication of small numbers is through the memorization of a multiplication table. This allows a carpenter to perform mental calculations on the go without the need of paper or a calculator.

Multiplication Table

	0	1	2	3	4	5	6	7	8	9	10	11	12
0	0	0	0	0	0	0	0	0	0	0	0	0	0
1	0	1	2	3	4	5	6	7	8	9	10	11	12
2	0	2	4	6	8	10	12	14	16	18	20	22	24
3	0	3	6	9	12	15	18	21	24	27	30	33	36
4	0	4	8	12	16	20	24	28	32	36	40	44	48
5	0	5	10	15	20	25	30	35	40	45	50	55	60
6	0	6	12	18	24	30	36	42	48	54	60	66	72
7	0	7	14	21	28	35	42	49	56	63	70	77	84
8	0	8	16	24	32	40	48	56	64	72	80	88	96
9	0	9	18	27	36	45	54	63	72	81	90	99	108
10	0	10	20	30	40	50	60	70	80	90	100	110	120
11	0	11	22	33	44	55	66	77	88	99	110	121	132
12	0	12	24	36	48	60	72	84	96	108	120	132	144

Goodheart-Willcox Publisher

Method Used to Multiply Whole Numbers

When multiplying numbers, they can be set up and multiplied either horizontally or vertically. Using the smoke detector example, the numbers can be multiplied:

Horizontal multiplication:

$$4 \times 5 = 20$$

Vertical multiplication:

$$\begin{array}{r} 4 \\ \underline{\times 5} \\ 20 \end{array}$$

Multiplication can be written horizontally for smaller numbers that can be multiplied by memory. Vertical multiplication is the preferred method on larger numbers because they must be worked out in steps. Multiplication problems are always set up with the numbers aligned on the right side with equivalent place values of each number lining up with each other. The number with the most digits is generally placed on top in the operation.

Example 4-1

$$\begin{array}{r} 143 \\ \times\ 12 \\ \hline \end{array}$$

Working right to left, multiply every digit in the top number (143) by the digit in the ones column of the bottom number (2). The answer (286) is placed below the equals line with the right digit lined up with the multiplier (2).

$(2 \times 3 = 6)$ $(2 \times 4 = 8)$ $(2 \times 1 = 2)$

$$\begin{array}{r} 143 \\ \times\ 12 \\ \hline 286 \end{array}$$

Working right to left, multiply each digit in the top number (143) by the digit in the tens column of the bottom number (1). The 1 is in the tens column, so place a 0 in the ones column of the product.

$(1 \times 3 = 3)$ $(1 \times 4 = 4)$ $(1 \times 1 = 1)$

$$\begin{array}{r} 143 \\ \times\ 12 \\ \hline 286 \\ 1{,}430 \end{array}$$

Add the products from each of the multiplication steps to obtain the final answer.

$$\begin{array}{r} 143 \\ \times\ 12 \\ \hline 286 \\ +\ 1{,}430 \\ \hline 1{,}716 \end{array}$$

Example 4-2

The same steps are used when multiplying larger numbers.

$$\begin{array}{r} 2{,}331 \\ \times\ 131 \\ \hline 2{,}331 \\ 69{,}930 \\ 00 \end{array}$$

When multiplying by the hundreds column in the second number, two zeros must now be placed when calculating the product. The 1 in the hundreds column is actually 100.

$$\begin{array}{r} 2{,}331 \\ \times\ 131 \\ \hline 2{,}331 \\ 69{,}930 \\ +\ 233{,}100 \\ \hline 305{,}361 \end{array}$$

Math Tip

When multiplying numbers, be sure to add the correct number of zeros to the products when working with numbers with multiple digits. Forgetting to do this will lead to confusion and inaccurate calculations.

Carrying Units in Multiplication Problems

As in addition, numbers may need to be carried over when multiplying larger numbers. The same rule applies; only one digit may occupy a column.

Example 4-3

In the following example, begin by multiplying ($3 \times 5 = 15$). The 5 is placed in the ones column and the 1 is carried and placed directly above the 7 in the tens column.

$$\begin{array}{r} {\scriptstyle 1} \\ 375 \\ \times\ 43 \\ \hline 5 \end{array}$$

Now, multiply ($3 \times 7 = 21$) and add the carried 1, making the answer 22. Place a 2 in the tens column and carry the other 2 by directly placing it over the 3 in the hundreds column.

$$\begin{array}{r} {\scriptstyle 2\,1} \\ 375 \\ \times\ 43 \\ \hline 25 \end{array}$$

Finally, multiply ($3 \times 3 = 9$) and add the carried 2, making the answer 11.

$$\begin{array}{r} {\scriptstyle 2\,1} \\ 375 \\ \times\ 43 \\ \hline 1{,}125 \end{array}$$

The next step is to multiply the top number (375) by the number in the tens column (4). It is important here to disregard the numbers that were carried over from the previous operations and carry new numbers as needed.

Begin here by multiplying ($4 \times 5 = 20$). The 0 is placed in the tens column and the 2 is carried and placed directly above the 7 in the tens column.

$$\begin{array}{r} {\scriptstyle 2} \\ 375 \\ \times\ 43 \\ \hline 1{,}125 \\ 00 \end{array}$$

(Continued)

Now, multiply (4 × 7 = 28) and add the carried 2 making the answer 30. Place a 0 in the tens column and carry the 3 by directly placing it over the 3 in the hundreds column.

$$\begin{array}{r} {\scriptstyle 3\,2} \\ 375 \\ \times\ 43 \\ \hline 1{,}125 \\ 000 \end{array}$$

Next, multiply (4 × 3 = 12) and add the carried 3, making the answer 15.

$$\begin{array}{r} {\scriptstyle 3\,2} \\ 375 \\ \times\ 43 \\ \hline 1{,}125 \\ 15{,}000 \end{array}$$

Finally, add the numbers from each of the multiplication steps to obtain the final answer.

$$\begin{array}{r} {\scriptstyle 3\,2} \\ 375 \\ \times\ 43 \\ \hline 1{,}125 \\ +\ 15{,}000 \\ \hline 16{,}125 \end{array}$$

Example 4-4

If it takes 48 receptacle boxes to wire a spec home, how many boxes will be needed to wire 14 more identical homes?

$$\begin{array}{r} 48 \\ \times 14 \\ \hline 192 \\ +\ 480 \\ \hline 672 \end{array}$$

Checking Multiplication

Checking for accuracy when multiplying can be done by two methods. The first method is to simply reverse the position of the numbers in the problem and repeat the multiplication. This process will produce different numbers from multiplying each digit as the problem is worked, but the numbers will add up to the same product if both problems were performed accurately. A mistake can easily be made in multiplication and found when the problem is reversed and checked.

Original Problem Multiplied Incorrectly	Reversal Check Multiplied Correctly
71	4 16
× 16	× 71
426	16
+ 71	+ 1,120
479	1,136

Multiplication can also be checked by using division. This will be discussed in the next unit.

Multiplying Denominate Numbers

The rules for multiplying denominate numbers differ from addition or subtraction and are applied based on the following:

Rule #1: When multiplying a denominate number by a whole number, the denominate numbers unit of measure is applied to the answer.

$$12 \times 8' = 96'$$

Rule #2: When multiplying more than one denominate number, they must have the same type of unit measure if possible. For example, when multiplying two lengths, both measurements should use the same units. The unit of measure is then squared. A subfloor measures 28′ wide × 52′ long. When multiplied together, the floor is said to contain 1,456 square feet or 1,456 sq ft.

$$28' \times 52' = 1{,}456 \text{ sq ft}$$

Carpentry Notes

Calculate Door Casing

*See Appendix A
Interior Finish

When installing trim in a new home under construction, a carpenter is tasked with installing ***casing**** on 22 interior doors. Prior to installation, the carpenter will need to order the casing for the doors. Each door requires 5 pieces of casing. Calculating a total is determined through multiplication.

$$22 \times 5 = 110 \text{ pieces of casing}$$

Carpentry Notes

Calculate Shingles and Siding

When it comes to calculating the amounts of shingles or siding, it is important to know that they are calculated and purchased by the **square**. A square is defined as 100 square feet. To order siding for a wall measuring 12′ × 32′, you would first multiply the denominate numbers and then divide the answer by 100 to determine the number of squares you would need.

$$12' \times 32' = 384 \text{ sq ft}$$

384 sq ft divided by 100 = 3.84 squares of siding needed

Unit 4 Review

Name ____________________ Date __________ Class __________

Multiply the following numbers without the use of a calculator. Show all of your work.

1. 5×7

2. 2×8

3. 4×4

4. 8×5

5. 7×3

6. 15×7

7. 24×8

8. 31×6

9. 98×4

10. 42×5

11. 24×18

12. 68×34

13. 28×21

14. 63×53

15. 74×27

Multiply the following numbers without the use of a calculator and then check for accuracy by reversing the position of the numbers in the problem. Show all your work.

16. $\begin{array}{r} 648 \\ \times\ 47 \\ \hline \end{array}$

17. $\begin{array}{r} 288 \\ \times\ 95 \\ \hline \end{array}$

18. $\begin{array}{r} 724 \\ \times\ 327 \\ \hline \end{array}$

19. $\begin{array}{r} 555 \\ \times\ 735 \\ \hline \end{array}$

20. $\begin{array}{r} 489 \\ \times\ 318 \\ \hline \end{array}$

21. $\begin{array}{r} 766 \\ \times\ 564 \\ \hline \end{array}$

22. $\begin{array}{r} 4{,}951 \\ \times\ \ \ 54 \\ \hline \end{array}$

23. $\begin{array}{r} 1{,}432 \\ \times\ \ 673 \\ \hline \end{array}$

24. $\begin{array}{r} 2{,}681 \\ \times\ 2{,}916 \\ \hline \end{array}$

25. $\begin{array}{r} 8{,}342 \\ \times\ 1{,}768 \\ \hline \end{array}$

26. A roofing crew is installing 3 squares of shingles per hour. At this rate, how many squares will they install in an 8-hour day?

____

Name ______________________ Date ____________ Class ____________

27. What is the total cost for 18 squares of vinyl siding that sells for $80 per square?

____

28. A business that sells garden sheds profits $225 per shed. What would their total profits be from selling 24 sheds?

____

29. A contractor is ordering ***gutters**** for a house. How many total feet of gutters need to be ordered for two sections 18′ long, three sections 24′ long, and one section 12′ long?

____

***See Appendix A**

Exterior Finish

30. A contractor needs 5 pieces of casing to trim interior doors and 2 pieces of casing to trim windows. The house has 14 doors and 17 windows. How many pieces of casing are needed to complete this job?

____

31. To order shingles for the home shown in the following illustration, determine the square feet of roof surface for the entire house.

____

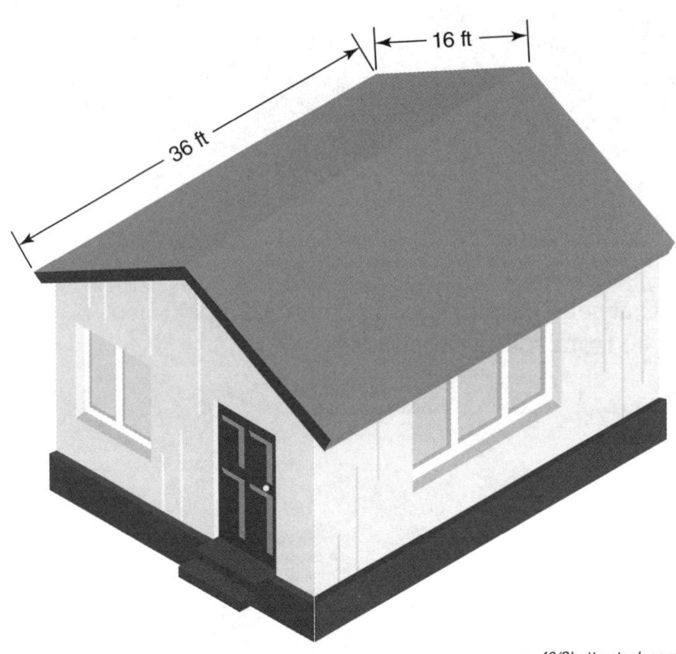

ps-42/Shutterstock.com

32. A worker needs to apply primer in 12 classrooms each with 220 sq ft of wall space. What is the total square feet of wall space to be primed?

____

Name _______________ Date _______________ Class _______________

33. A truss company needs 14 gang plates (7 per side) to manufacture the truss shown in the following image. They currently have an order for 36 trusses. How many gang plates are needed to fill this order?

____

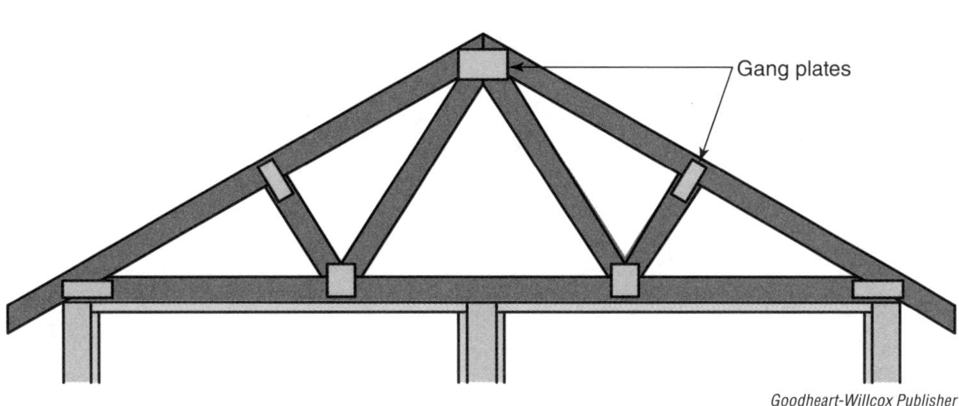

Goodheart-Willcox Publisher

34. If there are 26 three-tab shingles in a bundle, how many tabs will 7 bundles provide if three tabs are cut from one shingle?

____

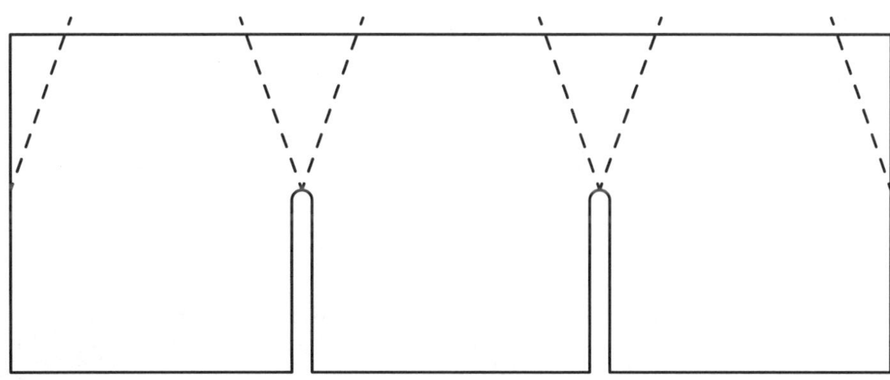

Goodheart-Willcox Publisher

35. A contractor's foreman worked 52 hours at $22 an hour. Three carpenters worked 46 hours each at $16 an hour. One helper worked 44 hours at $9 an hour. What was the total the contractor paid in wages to this crew?

____

UNIT 5

Dividing Whole Numbers

Objectives

After studying this unit, you will be able to:

- Practice the method used to divide whole numbers.
- Solve division problems with remainders.
- Check division problems with multiplication.
- Solve division problems with denominate numbers.

Division is the last of the four basic mathematical operations. While subtraction was thought of as the inverse of addition, division is thought of as the inverse of multiplication. **Division** is the operation of separating a quantity into a specific number of equally sized portions. For example, if you needed to know how many times 7 can go into 84, this can be determined by subtracting 7 multiple times. After multiple steps of subtraction problems, we determine that 7 can go into 84 a total of 12 times:

(1) $84 - 7 = 77$
(2) $77 - 7 = 70$
(3) $70 - 7 = 63$
(4) $63 - 7 = 56$
(5) $56 - 7 = 49$
(6) $49 - 7 = 42$
(7) $42 - 7 = 35$
(8) $35 - 7 = 28$
(9) $28 - 7 = 21$
(10) $21 - 7 = 14$
(11) $14 - 7 = 7$
(12) $7 - 7 = 0$

Using division, this problem can be solved much faster and with fewer steps. A **division sign (÷)** is placed between the numbers to signify a division problem.

Using division, this problem can be solved much faster and with fewer steps.

$$84 \div 7 = 12$$

This division problem can be expressed in multiple ways:

- 84 divided by 7 equals 12.
- There are 12 7s in 84.
- 12 goes into 84 7 times.
- 7 divides into 84 12 times.

Example 5-1

Division is useful in determining quantities of materials or performing estimates. A carpenter uses division to determine how many sheets of subfloor to buy to cover a 40′ × 36′ floor system.

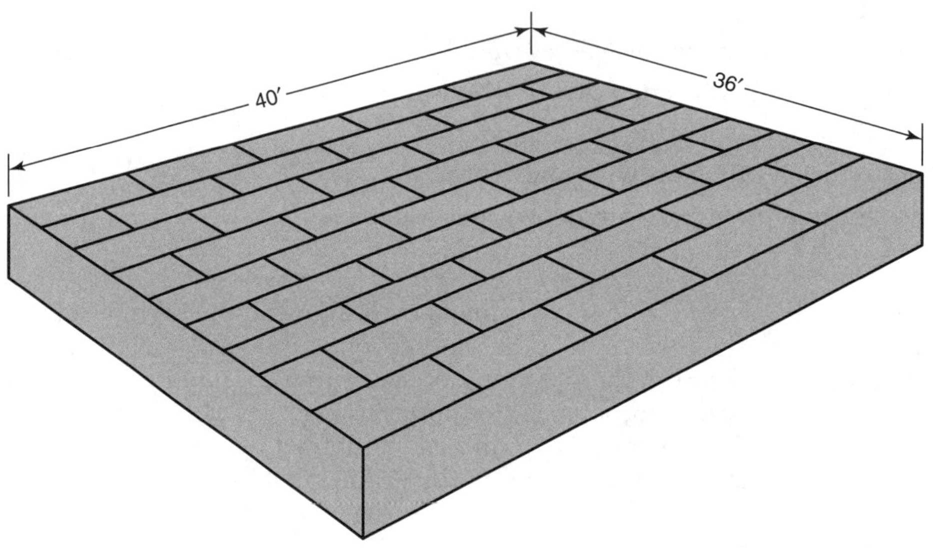

Goodheart-Willcox Publisher

Multiply the length of the floor system by the width to determine the area:

$$40' \times 36' = 1{,}440 \text{ sq ft}$$

Divide the total by 32 since 1 sheet will cover 32 sq ft:

$$1{,}440 \text{ sq ft} \div 32 \text{ sq ft/sheet} = 45 \text{ sheets}$$

Example 5-2

*See Appendix A
Interior Finish

How many pieces of 8′ ***base**** trim would be needed to cover a 40′ × 36′ room?

Determine the perimeter of the room:

$$40' + 40' + 36' + 36' = 152'$$

Divide the perimeter by 8, since each piece of trim measures 8′ long.

$$152' \div 8' = 19 \text{ pieces of trim}$$

Method Used to Divide Whole Numbers

While addition (+), subtraction (–), and multiplication (×) each had a symbol to identify their operations, division can be identified by four symbols and methods to set up the problem.

Division bracket $7\overline{)84}$

Division sign $84 \div 7$

Division bar $\frac{84}{7}$

Slanted bar $84/7$

Because division problems can be set up using multiple symbols, it is important to learn the terminology and definitions associated with a common division problem.

$$\begin{array}{r} 12 \leftarrow \text{Quotient} \\ \text{Divisor} \rightarrow 7\overline{)84} \leftarrow \text{Dividend} \end{array}$$

The **dividend** is the number that is being divided into smaller parts by the divisor.

The **divisor** is the number by which the dividend is being divided.

The **quotient** is the number of times the dividend is divided by the divisor.

While it is easy to recognize that $24 \div 12 = 2$, calculators are needed on many division problems because of their complexity. When working on a jobsite where a calculator may not be available, the long division bracket method can be used to solve complex division problems.

Example 5-3

Set up the problem by placing the dividend and divisor in their correct locations. For this example, division brackets are used.

$$8\overline{)1{,}968}$$

Starting on the left of 1,968, determine how many times the divisor (8) can go into the first digit of the dividend (1,968). In this example, 8 is larger than 1, so it cannot. When this happens, you then move one more digit to the right on the dividend and try again. This time 8 can divide into 19 two times. Since we were dividing into 19, the 2 is written directly above the 9 in 19.

$$\begin{array}{r} 2 \\ 8\overline{)1{,}968} \end{array}$$

After placing the 2 above the 9, multiply 2×8 and place the answer (16) under the 19. Subtract and insert the 3 as the answer.

$$\begin{array}{r} 2 \\ 8\overline{)1{,}968} \\ -1\,6 \\ \hline 3 \end{array}$$

(Continued)

Bring down the next digit in the dividend (6) and place it to the right of the 3. The new remainder becomes 36. It is very important when bringing down numbers that they move directly down and remain in the same column.

$$\begin{array}{r} 2 \\ 8\overline{)1{,}968} \\ -16\downarrow \\ \hline 36 \end{array}$$

The division process continues by determining if 8 will divide into 36. In this case it does, 4 times. The 4 is placed above the 6 and multiplied by the divisor (4 × 8). The answer (32) is placed below the 36 and subtracted.

$$\begin{array}{r} 24 \\ 8\overline{)1{,}968} \\ -16 \\ \hline 36 \\ -32 \\ \hline 4 \end{array}$$

This cycle is repeated until all the digits in the dividend have been used.

$$\begin{array}{r} 246 \\ 8\overline{)1{,}968} \\ -16 \\ \hline 36 \\ -32\downarrow \\ \hline 48 \\ -48 \\ \hline 0 \end{array}$$

Math Tip

Neat and orderly handwriting is very important when calculating math operations. Keeping numbers properly aligned when dividing will help to avoid errors. Misaligned numbers can make it easy to bring down a number from the wrong column.

Remainders in Division Problems

The type of division problems covered so far in this unit have been ones in which the divisor divides evenly into the dividend.

In reality, many division problems do not work out evenly. A **remainder** is a number that remains after the entire dividend has been divided. The remainder is attached to the quotient with the letter "r."

Example 5-4

$$\begin{array}{r} 106\text{ r}3 \\ 9\overline{)957} \\ -9 \\ \hline 05 \\ -0 \\ \hline 57 \\ -54 \\ \hline 3 \end{array}$$

Checking Division

Division can be thought of as the inverse of multiplication. Because of this, division problems can be checked by multiplication. Once a division problem has been calculated, it can be checked by multiplying the quotient by the divisor. If there was a remainder in the answer, it is then added to the answer. For example, the division problem 76 ÷ 6 is checked with multiplication.

$$\begin{array}{r} 12\text{ r}4 \\ 6\overline{)76} \\ -6 \\ \hline 16 \\ -12 \\ \hline 4 \end{array} \qquad \begin{array}{r} \scriptstyle 1 \\ 12 \\ \times\ 6 \\ \hline 72 \\ +\ 4 \\ \hline 76 \end{array}$$

Division can also be used to check the accuracy of a multiplication problem.

$$48 \div 6 = 8 \text{ or } 48 \div 8 = 6$$

Dividing Denominate Numbers

When dividing a denominate number by a whole number, the denominate number's unit of measure is applied to the final answer.

$$12 \text{ yd} \div 3 = 4 \text{ yd}$$

$$18' \div 6 = 3'$$

When dividing two denominate numbers containing a squared dividend, the divisor's unit of measure will cancel out the part of the dividend's unit they have in common, resulting in the answer containing a single unit of measure.

$$\frac{81 \text{ sq ft}}{9'} = 9'$$

Carpentry Notes

Equal Spacing of Posts

Division is an operation that is often used to space objects out equally. For example, during the erection of a fence, locations need to be established and drilled for the fence posts. The 56′ fence is built with 9 equally spaced poles. Division is used to determine the post spacing.

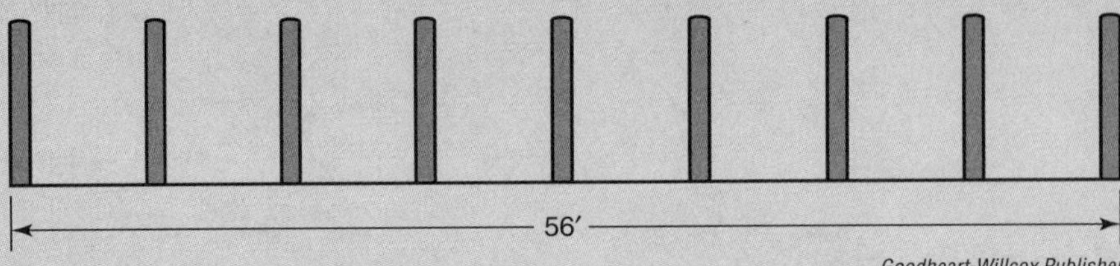

Goodheart-Willcox Publisher

Since there will be 8 equal spaces between the 9 poles, the 56′ length will be divided by 8.

$$56' \div 8 = 7'$$

Through division, it has been determined that spacing from post to post will be 7′.

Unit 5 Review

Name ______________________ Date ____________ Class ____________

Divide the following numbers without the use of a calculator. Show all of your work.

1. $35 \div 7 =$

2. $12 \div 3 =$

3. $18 \div 6 =$

4. $24 \div 4 =$

5. $45 \div 9 =$

6. $6\overline{)210}$

7. $4\overline{)344}$

8. $11\overline{)588}$

9. $8\overline{)4{,}975}$

10. $13\overline{)3{,}627}$

11. $18\overline{)15{,}148}$

12. $127\overline{)2{,}132}$

13. $352\overline{)44{,}444}$

14. $249\overline{)170{,}316}$

Name ______________________________ **Date** ____________ **Class** ____________

Divide the following numbers without the use of a calculator. Then, check your accuracy through multiplication. Show all your work.

15. 3,465 ÷ 55 =

16. 2,106 ÷ 27 =

17. Divide 8,364 by 68.

18. 6,315 ÷ 145 =

19. 46,972 ÷ 264 =

20. 7,461 divided by 37

21. How long does it take to install 450 sq ft of hardwood flooring at the rate of 75 sq ft per hour?

____

22. A room measures 14′ × 16′ and has 8′ high walls. At 400 sq ft of coverage per gallon, how many gallons are needed to paint the walls with two coats of paint?

____

23. How many 4′ pieces can be cut from a 16′ board?

____

24. A worker was paid $1,280 for 80 hours of work. How much did the worker earn per hour?

____

Name ______________________ Date ____________ Class ____________

25. A 72′ long pole barn needs trusses installed every 2′. How many trusses, including the trusses at the front and back of the 72′ barn, need to be ordered?

____

Use the following image of a room to answer problems 26–28.

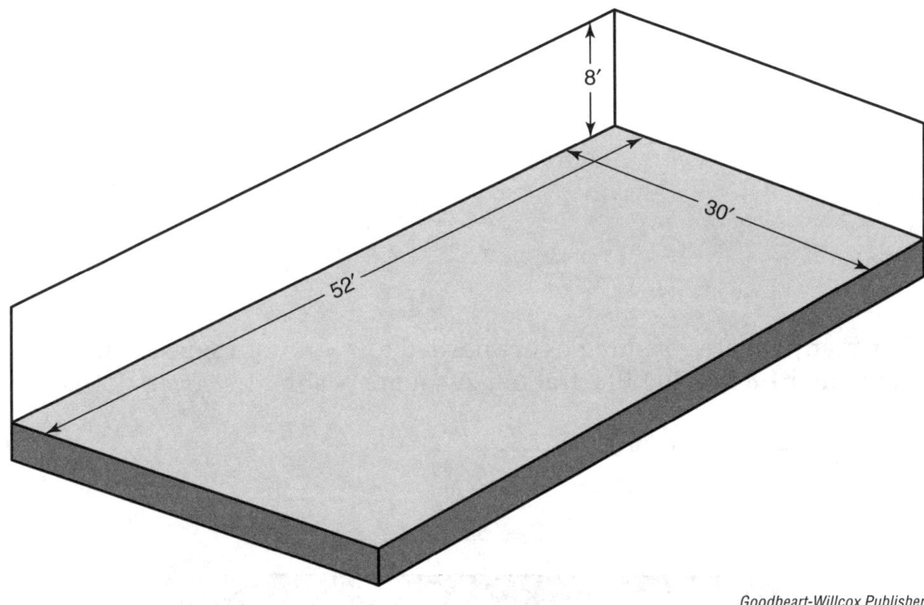

Goodheart-Willcox Publisher

26. How many sheets of subflooring are needed to cover the floor system, when each sheet covers 32 sq ft?

____

*See Appendix A
Wall Frame

27. To build the four exterior walls for the house, the builder is using $2 \times 4 \times 16'$ material for the ***sole plates**** and single ***top plates****. How many 16′ boards are needed for plates on the four exterior walls?

____

28. How many pieces of sheeting are needed to cover the exterior walls, when each sheet covers 32 sq ft?

____

29. How many squares of shingles are needed to cover both sides of the roof shown in the following illustration? A square of shingles covers 100 sq ft.

____

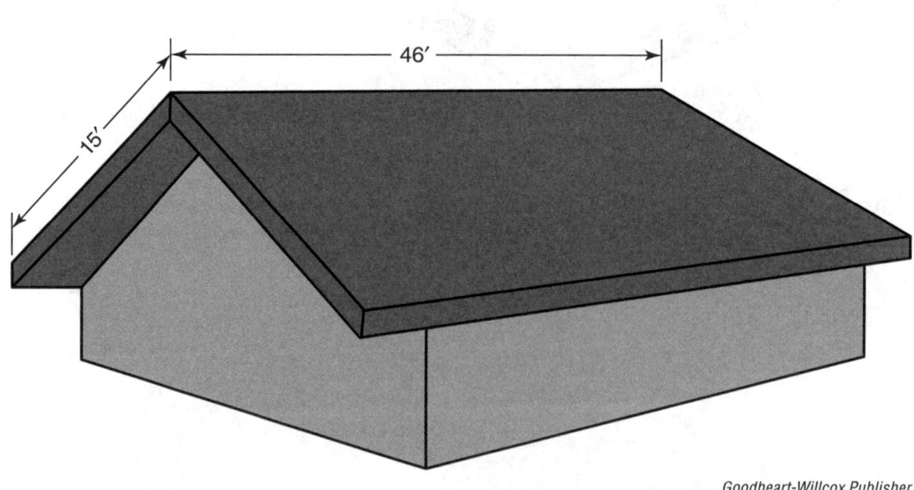

Goodheart-Willcox Publisher

Name ______________________ Date ____________ Class ____________

30. How many studs are needed to frame a 32′ wall (no windows or doors)? Studs will be placed 16″ on center (16″ between the centers of each stud).

____

31. How many 2′ × 4′ ceiling tiles are needed for a 28′ × 56′ suspended ceiling?

____

Use the following staircase plan to answer questions 32–33.

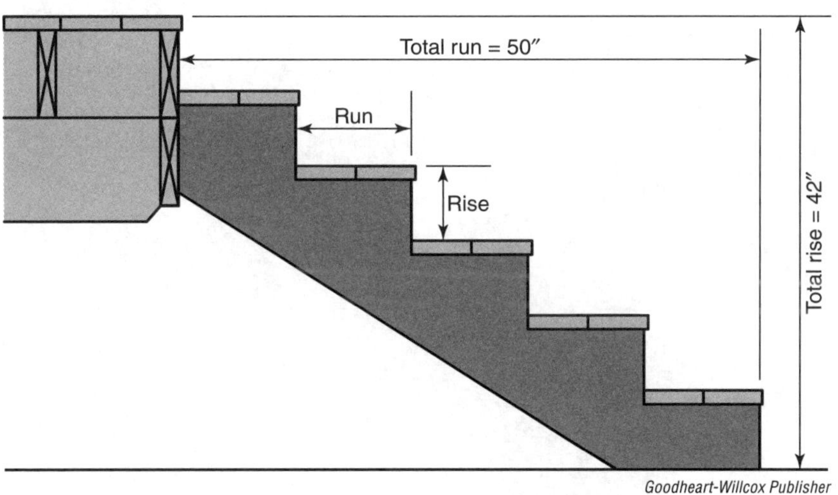

32. What is the rise (height) of each step, assuming uniform rise in each step?

____

33. What is the run (width) of each step, assuming uniform run in each step?

____

34. How many rolls of insulation are needed to insulate 1,680 sq ft of walls if each roll contains 49 sq ft of insulation?

____

35. A contractor bidding a project has allowed $1,620 for labor. How many total hours can his 3-person crew work and still remain under budget, when each worker is making $12/hour?

____

Section 1 Exam

Name ______________________________ Date ____________ Class ____________

Use the number 6,381,947 to answer questions 1–3.

1. Which digit is in the hundreds place?

   ____

2. Which digit is in the ten thousands place?

   ____

3. In which place value is the 3?

   ____

4. Round 319,462 to the thousands place value.

   ____

5. Three carpenters cut 16 deck posts to a length of 67″. Identify the denominate number.

   ____

6. 47″ + 52″ + 143″ =

7. 73 sq ft + 39 sq ft + 177 sq ft + 348 sq ft =

8. 458 yd + 145 yd + 227 yd + 98 yd =

9.
```
  323 sq ft
– 268 sq ft
```

10. $\begin{array}{r} 42{,}291 \\ -\ 1{,}788 \\ \hline \end{array}$

11. $\begin{array}{r} 14{,}500 \text{ bd ft} \\ -\ 9{,}788 \text{ bd ft} \\ \hline \end{array}$

12. $\begin{array}{r} 15 \\ \times\ 9 \\ \hline \end{array}$

13. $\begin{array}{r} 42 \\ \times 17 \\ \hline \end{array}$

14. $\begin{array}{r} 438 \\ \times 612 \\ \hline \end{array}$

15. $\begin{array}{r} 3{,}487 \\ \times\ \ 525 \\ \hline \end{array}$

Name ______________________________ Date ____________ Class ____________

16. $1,566 \div 18 =$

17. $3,971 \div 76 =$

18. Divide 5,264 by 72.

19. $36,157 \div 444 =$

20. A carpenter is giving a quote to install baseboard trim in rooms whose perimeters measure 52, 64, 48, 72, and 28 lineal feet. How many total lineal feet of trim is required?

____

21. A crew of 4 construction workers worked three 11-hour days, one 7-hour day, and one 9-hour day. How many total hours were worked by the crew?

____

22. A contractor bid $45,000 on a construction project. Once the project was complete, $18,721 was spent on materials, $7,687 was paid to his workers, and $4,600 paid to a subcontractor. How much was left over for the carpenter to claim as profit?

____

23. If three 18″ boards, two 9″ boards, and one 17″ board were cut from a 168″ board, how much is left over?

____

24. A carpenter is installing wooden shelving in all the closets in a new home. Two closets need 7′ shelves, six closets need 6′ shelves, a linen closet needs five 2′ shelves, and the pantry needs six 4′ shelves. How many lineal feet of shelving is needed?

____

25. A contractor budgeted $1,800 for labor for a job. How many hours can two workers, each making $14 an hour, work and remain under budget?

____

SECTION 2

Fractions

Key Terms

cancel
common denominator
denominator
equivalent fraction
fraction
fraction bar
higher terms
improper fraction
lowest common denominator (LCD)
lowest terms
mixed number
numerator
prime factors
prime number
proper fraction
reciprocal
reducing

UNIT 6

Basic Principles of Fractions

Objectives

After studying this unit, you will be able to:

- Identify the parts of a fraction.
- Recognize proper and improper fractions.
- Demonstrate how to create equivalent fractions.
- Demonstrate how to reduce fractions to their lowest terms using prime factors.
- Demonstrate how to convert mixed numbers to improper fractions.
- Demonstrate how to find the lowest common denominator (LCD).

Fractions are used every day by carpenters when reading a tape measure (5 1/4"), when expressing time (1 1/2 hours), the volume of concrete (8 1/3 cu yd), and how many squares of shingles are needed to cover the area of a gable roof (17 2/3 squares). **Fractions** are used to represent a part of a whole unit. It is important for a carpenter to understand the basic principles of fractions before beginning to measure, add, subtract, multiply, or divide them.

Parts of a Fraction

Fractions are made up two numbers separated by a **fraction bar**. The bar can be horizontal or slanted and has the same meaning as the ÷ sign.

$$\frac{3}{4} = 3/4$$

The following ruler reads 9/16 of an inch. The whole inch is divided into 16 equal parts. This total division number (16) becomes the **denominator** in the fraction, while the 9 highlighted parts become the **numerator**.

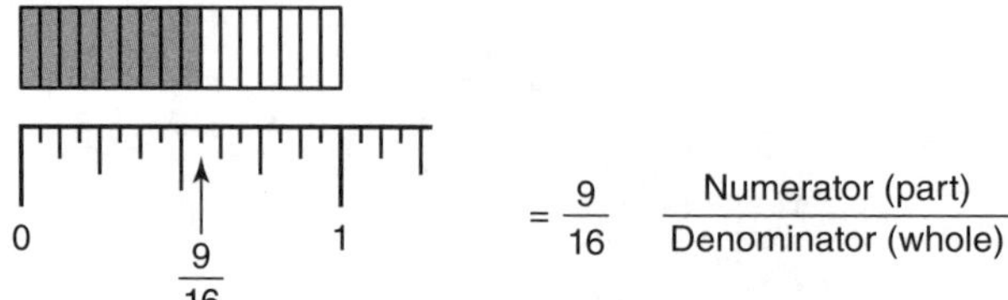

Goodheart-Willcox Publisher

Example 6-1

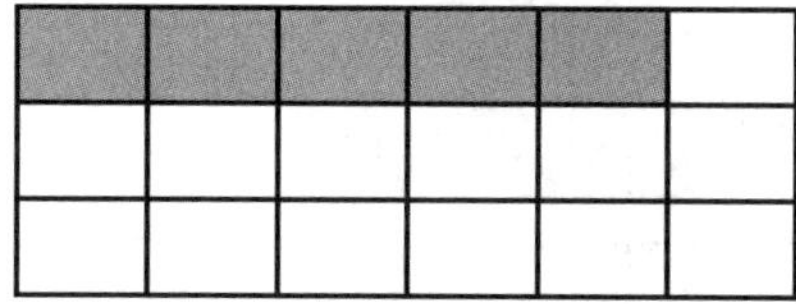

Goodheart-Willcox Publisher

To determine the fractional value of the highlighted parts shown, start by determining the number of parts in the whole. In this example, the whole is divided into 18 equal parts. The denominator becomes 18, representing the number of parts in the whole.

The numerator is the number of parts highlighted. In this example, 5 parts of the whole are highlighted and becomes the top number of the fraction.

The fractional value is 5/18.

Proper and Improper Fractions

A **proper fraction** is one in which the numerator is less than the denominator and the total value is less than one. Examples of proper fractions are:

$$\frac{3}{8} \qquad \frac{1}{2} \qquad \frac{3}{4} \qquad \frac{7}{16}$$

An **improper fraction** is one in which the numerator is greater than the denominator and their value is greater than one. Examples of improper fractions are:

$$\frac{12}{5} \qquad \frac{17}{4} \qquad \frac{47}{8} \qquad \frac{9}{2}$$

Math Tip

A fraction having the same numerator and denominator is equivalent to one.

$$\frac{8}{8} = 1$$

Equivalent Fractions

Equivalent fractions are fractions that have the same value but with different numerators and denominators. Fractions are used to show the relationship of a part to a whole. The same relationship can be expressed in multiple ways.

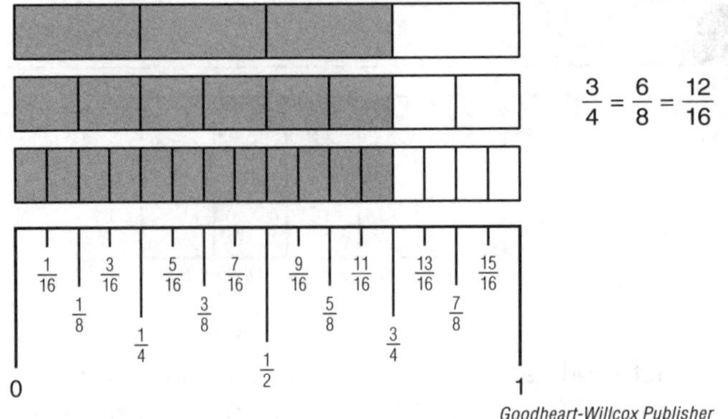

Goodheart-Willcox Publisher

Fractions can be changed to produce an equivalent fraction in **higher terms**. This is achieved by multiplication of fractions, which will be discussed in more detail in Unit 9, *Multiplying Fractions*. To produce a fraction expressed in higher terms, the original fraction is multiplied by a fraction with the same numerator and denominator (a fraction equal to one). The numerators are multiplied and placed as the numerator of the product. Then, the denominators are multiplied and placed as the denominator of the product. The resulting product is equivalent to the original fraction in higher terms.

$$\frac{3}{4} \times \frac{2}{2} = \frac{6}{8}$$

Reducing Fractions to Their Lowest Terms

A fraction is considered to be in its **lowest terms** when it has the lowest possible numerator and denominator.

The fractions 1/3, 2/5, 7/13 are in their lowest terms.

The fractions 6/27, 8/30, 4/10 are not in lowest terms and can be reduced.

When fractions are not in their lowest terms, **reducing** is used to convert the fraction into its lowest terms. This involves finding numbers that divide evenly into the numerator and the denominator. Continue breaking down the numerator and denominator into numbers they can easily be divided into until there are only prime numbers. A **prime number** is a number that can be divided evenly by only 1 and itself, such as 2, 3, 5, and 7. Determining all of the prime numbers that a larger number can be divided into it gives you a list of that number's **prime factors**. Prime factors, when multiplied together, will give you the original, larger number.

Example 6-2

To determine the prime factors of 24, start with the lowest prime number that will divide evenly. In this case it is 2.

$24 \div \mathbf{2} = 12$ 12 is not a prime number; divide 12 by 2.

$12 \div \mathbf{2} = 6$ 6 is not a prime number; divide 6 by 2.

$6 \div \mathbf{2} = \mathbf{3}$ 3 is a prime number.

The prime factors of 24 are $2 \times 2 \times 2 \times 3$.

When using prime factors to reduce a fraction, determine the prime factors of both the numerator and the denominator. Write the fraction with the prime factors and cross out factors that the numerator has in common with the denominator.

Example 6-3

To reduce 36/98 to its lowest terms, start by determining the prime factors of the numerator.

$$36 \div \mathbf{2} = 18$$
$$18 \div \mathbf{2} = 9$$
$$9 \div \mathbf{3} = \mathbf{3}$$

The prime factors of 36 are $2 \times 2 \times 3 \times 3$.

Determine the prime factors of the denominator 98.

$$98 \div \mathbf{2} = 49$$
$$49 \div \mathbf{7} = \mathbf{7}$$

The prime factors for 98 are $2 \times 7 \times 7$.

Rewrite the fraction with the prime factors. Cross out common factors that both numbers can be divided by. The remaining prime factors will show the reduced fraction in its lowest terms.

$$\frac{36}{98} = \frac{\cancel{2} \times 2 \times 3 \times 3}{\cancel{2} \times 7 \times 7} = \frac{18}{49}$$

Math Tip

Common denominators are needed when performing certain math calculations. This may require you to convert a fraction into a higher term to complete the problem. After the calculations are completed, the final answer is reduced to its lowest term, if needed.

$$\frac{3}{8} + \frac{1}{4} = \frac{3}{8} + \boxed{\frac{2}{8}} = \frac{5}{8}$$

Converting Mixed Numbers to Improper Fractions

Mixed numbers consist of a whole number combined with a fraction. It is much simpler to visualize the value of a number as a mixed number, rather than an improper fraction.

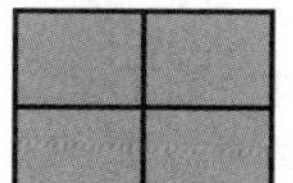
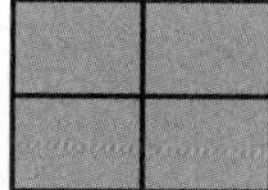
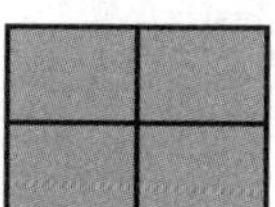
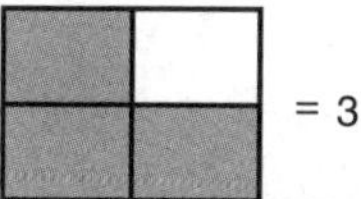

Goodheart-Willcox Publisher

When performing math calculations, it is sometimes necessary to convert mixed numbers into an improper fraction. Improper fractions can be simplified by dividing the numerator by the denominator to produce either a mixed number or a whole number. Final answers should always be converted back to a mixed number to express them in their proper form. If the mixed number in the previous example is a denominate number, such as 3 3/4 cu yd, it would not be proper to express as 15/4 cu yd.

Example 6-4

Remember, the fraction bar has the same meaning as the division sign.

$$\frac{20}{4} = 20 \text{ divided by } 4 = 5$$

Math Tip

Any whole number can also be written as a fraction by placing it over a denominator of 1.

$$5 = \frac{5}{1}$$

Example 6-5

A mixed number results when there is a remainder left over from the division. The mixed number consists of the whole number that was divided and the remainder becomes the new numerator over the original denominator.

$$\frac{47}{12} = 47 \div 12 = 3 \text{ r}11 = 3\frac{11}{12}$$

When multiplying and dividing fractions, mixed numbers must first be converted to improper fractions.

Example 6-6

Three steps are used in the following example to change the mixed number 8 3/8 to an improper fraction.

1. Multiply the whole number times the denominator.

$$8 \times 8 = 64$$

2. Add the numerator to the answer from step 1.

$$64 + 3 = 67$$

3. Place the answer from step 2 over the original denominator.

$$\frac{67}{8}$$

Finding the Lowest Common Denominator

When adding and subtracting fractions, both operations require common denominators. Fractions with **common denominators** have the same number as their denominator.

$$\frac{7}{16} \qquad \frac{1}{16} \qquad \frac{3}{16}$$

When fractions do not have common denominators, one or more of the fractions must be converted to have common denominators. It can be useful to find the lowest common denominator (LCD). The **lowest common denominator (LCD)** is the smallest shared multiple of each denominator.

Example 6-7

The fractions below cannot be added in their present state because the denominators are not common.

$$\frac{3}{8} + \frac{5}{6} + \frac{7}{12} =$$

When denominators are not common, the multiples of each denominator are listed until the LCD is determined.

$8 \times 1 = 8$	$6 \times 1 = 6$	$12 \times 1 = 12$
$8 \times 2 = 16$	$6 \times 2 = 12$	$12 \times 2 = \textcircled{24}$
$8 \times 3 = \textcircled{24}$	$6 \times 3 = 18$	$12 \times 3 = 36$
$8 \times 4 = 32$	$6 \times 4 = \textcircled{24}$	$12 \times 4 = 48$

Using 24 as the LCD for all the fractions, the numerators will need to be changed to keep each fraction equivalent. This is done by multiplying the numerator and the denominator by the same number.

$$\frac{3}{8} = \frac{?}{24} \qquad \frac{3}{8} \times \frac{3}{3} = \frac{9}{24}$$

$$\frac{5}{6} = \frac{?}{24} \qquad \frac{5}{6} \times \frac{4}{4} = \frac{20}{24}$$

$$\frac{7}{12} = \frac{?}{24} \qquad \frac{7}{12} \times \frac{2}{2} = \frac{14}{24}$$

Carpentry Notes

Measuring with Fractions

In the construction trades, fractions are as common as nails, screws, and lumber. Imagine tape measures that only measure to the nearest inch. How could anything ever be measured, cut, or assembled with any degree of accuracy? If the board shown were measured with a simple inch rule, there would be no way to accurately convey that measurement to another worker without fractions on the rule.

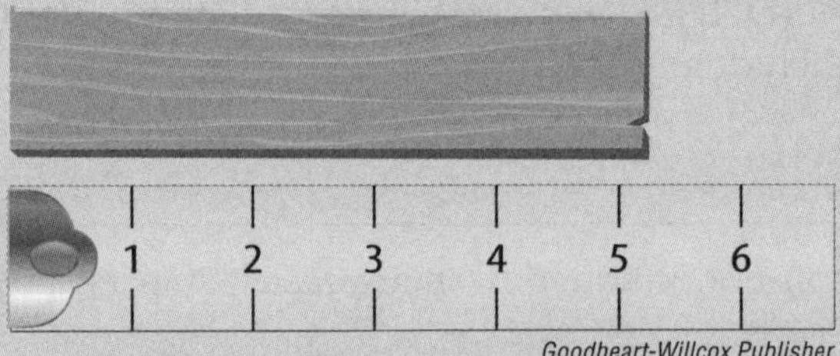

Goodheart-Willcox Publisher

Skilled craftsmen are known by their quality work. Fractions provide the degree of precision necessary to achieve accurate measurements and calculations within our trade.

Unit 6 Review

Name ______________________________ Date ____________ Class ____________

Complete the following as equivalent fractions.

1. $\frac{1}{8} = \frac{\quad}{32}$

2. $\frac{3}{4} = \frac{\quad}{64}$

3. $\frac{5}{12} = \frac{\quad}{96}$

4. $\frac{24}{27} = \frac{\quad}{9}$

5. $\frac{7}{16} = \frac{\quad}{128}$

6. $\frac{8}{96} = \frac{\quad}{12}$

7. $\frac{2}{2} = \frac{\quad}{22}$

8. $\frac{6}{13} = \frac{\quad}{65}$

9. $\frac{45}{162} = \frac{\quad}{18}$

10. $\frac{4}{14} = \frac{\quad}{168}$

Reduce the following fractions to their lowest term.

11. $\frac{5}{45} = —$

12. $\frac{4}{32} = —$

13. $\frac{27}{72} = —$

14. $\frac{6}{33} = —$

15. $\frac{24}{64} = —$

16. $\frac{55}{143} = —$

17. $\frac{88}{216} = —$

18. $\frac{6}{28} = —$

19. $\frac{132}{3520} = —$

20. $\frac{16}{480} = —$

Name ______________________ **Date** __________ **Class** __________

Convert the following improper fractions to mixed numbers and reduce to the lowest terms when necessary.

21. $\frac{14}{8} = —$

22. $\frac{21}{7} = —$

23. $\frac{29}{4} = —$

24. $\frac{24}{5} = —$

25. $\frac{52}{3} = —$

26. $\frac{139}{11} = —$

27. $\frac{32}{21} = —$

28. $\frac{44}{12} = —$

29. $\frac{45}{16} = —$

30. $\frac{65}{12} = —$

Convert the following mixed numbers to improper fractions.

31. $1\frac{5}{8} = \text{—}$

32. $5\frac{1}{4} = \text{—}$

33. $7\frac{3}{16} = \text{—}$

34. $2\frac{2}{9} = \text{—}$

35. $3\frac{6}{8} = \text{—}$

Determine the lowest common denominator (LCD) for the following pairs of fractions.

36. $\frac{3}{8}$ and $\frac{5}{12}$

37. $\frac{3}{4}$ and $\frac{2}{15}$

38. $\frac{5}{12}$ and $\frac{4}{5}$

39. $\frac{24}{8}$ and $\frac{2}{9}$

40. $\frac{7}{3}$ and $\frac{3}{28}$

UNIT 7

Adding Fractions

Objectives

After studying this unit, you will be able to:

- Demonstrate the method used to add fractions.
- Add fractions with common denominators.
- Add fractions without common denominators.
- Add mixed numbers.

A carpenter will use a tape measure to determine the length of a board. The length could be expressed in inches or feet but often will include a fraction. In construction, you will not only need to identify fractions by their value, but you will also need to add, subtract, multiply, and divide fractions quickly and accurately on a job site. There are many operations performed within the construction trades that will require carpenters to quickly add measurements with fractions.

Method Used to Add Fractions

When adding fractions, first determine what type of fractions are being used. Fractions may or may not have common denominators and can contain mixed numbers without common denominators. Each of these scenarios requires a different process to solve the problem, but all eventually lead to setting up an addition problem with common denominators.

As with adding whole numbers, fractions can be added horizontally or vertically.

Horizontal addition:

$$\frac{3}{16} + 5\frac{5}{8} = ?$$

Vertical addition:

$$\begin{array}{r} \frac{3}{16} \\ + 5\frac{5}{8} \\ \hline ? \end{array}$$

Math Tip

A general rule for adding fractions is that before any addition is begun, all the fractions must have common denominators.

Adding Fractions with Common Denominators

When adding fractions with common denominators, the numerators are added together and their sum is placed over the original denominator.

Example 7-1

$$\frac{\text{Numerators are added}}{\text{Denominator does not change}}$$

$$\frac{3}{16} + \frac{5}{16} = \frac{3+5}{16} = \frac{8}{16} = \frac{1}{2}$$

Math Tip

When performing any operation with fractions, always express the final answer in lowest terms.

Adding Fractions without Common Denominators

Fractions that do not have the same denominators cannot be added until the denominators are changed. The goal is to choose the lowest common denominator (LCD) for each of the original denominators. The process of determining the LCD is discussed in Unit 6, *Basic Principles of Fractions*.

Example 7-2

The fractions below cannot be added because the denominators are not the same, so the LCD must be determined.

$$\frac{1}{4} + \frac{3}{5} + \frac{7}{8} = ?$$

(Continued)

List the multiples of each denominator to determine the LCD.

$4 \times 1 = 4$	$5 \times 1 = 5$	$8 \times 1 = 8$
$4 \times 2 = 8$	$5 \times 2 = 10$	$8 \times 2 = 16$
$4 \times 3 = 12$	$5 \times 3 = 15$	$8 \times 3 = 24$
$4 \times 4 = 16$	$5 \times 4 = 20$	$8 \times 4 = 32$
$4 \times 5 = 20$	$5 \times 5 = 25$	$8 \times 5 = \textcircled{40}$
$4 \times 6 = 24$	$5 \times 6 = 30$	$8 \times 6 = 48$
$4 \times 7 = 28$	$5 \times 7 = 35$	$8 \times 7 = 56$
$4 \times 8 = 32$	$5 \times 8 = \textcircled{40}$	$8 \times 8 = 64$
$4 \times 9 = 36$	$5 \times 9 = 45$	$8 \times 9 = 72$
$4 \times 10 = \textcircled{40}$	$5 \times 10 = 50$	$8 \times 10 = 80$

$$\frac{1}{4} \times \frac{10}{10} = \frac{10}{40} \qquad \frac{3}{5} \times \frac{8}{8} = \frac{24}{40} \qquad \frac{7}{8} \times \frac{5}{5} = \frac{35}{40}$$

Once the LCD is found, each fraction is converted to an equivalent fraction using the LCD.

$$\frac{1}{4} + \frac{3}{5} + \frac{7}{8} = \frac{10}{40} + \frac{24}{40} + \frac{35}{40}$$

Finally, the fractions are added and reduced to the lowest terms.

$$\frac{10}{40} + \frac{24}{40} + \frac{35}{40} = \frac{69}{40} = 1\frac{29}{40}$$

Adding Mixed Numbers

Mixed numbers are added by determining the sum of the whole numbers and the fractions separately.

Example 7-3

$$\begin{array}{r} 12\frac{3}{16} \\ +\ 7\frac{5}{16} \\ \hline ? \end{array}$$

First, the fractions are added.

$$\frac{3}{16} + \frac{5}{16} = \frac{8}{16} = \frac{1}{2}$$

Then, the whole numbers are added.

$$12 + 7 = 19$$

$$19 + \frac{1}{2} = 19\frac{1}{2}$$

Mixed numbers without common denominators cannot be added until an LCD is determined.

Example 7-4

$$\begin{array}{r} 8\frac{3}{4} \\ +\,4\frac{2}{5} \\ \hline ? \end{array}$$

The whole numbers can easily be added, 8 + 4 = 12. The fractions cannot be added until the LCD is determined.

$$\begin{array}{lcl} 4 \times 1 = 4 & \qquad & 5 \times 1 = 5 \\ 4 \times 2 = 8 & & 5 \times 2 = 10 \\ 4 \times 3 = 12 & & 5 \times 3 = 15 \\ 4 \times 4 = 16 & & 5 \times 4 = \textcircled{20} \\ 4 \times 5 = \textcircled{20} & & 5 \times 5 = 25 \end{array}$$

$$\frac{3}{4} \times \frac{5}{5} = \frac{15}{20} \qquad \frac{2}{5} \times \frac{4}{4} = \frac{8}{20}$$

$$\begin{array}{r} 8\frac{15}{20} \\ +\,4\frac{8}{20} \\ \hline 12\frac{23}{20} \end{array}$$

Once the LCD is determined and the fractions are added, the answer contains an improper fraction. The fraction is then converted to a mixed number and the whole numbers are combined.

$$\frac{23}{20} = 1\frac{3}{20}$$

$$12 + 1\frac{3}{20} = 13\frac{3}{20}$$

Unit 7 Review

Name ______________________ **Date** ____________ **Class** ____________

Solve the following equations and reduce the answer to lowest terms. Show all of your work.

1. $\frac{4}{8} + \frac{3}{8} =$

2. $\frac{1}{5} + \frac{2}{5} =$

3. $\frac{2}{15} + \frac{5}{15} + \frac{3}{15} =$

4. $\frac{3}{16} + \frac{3}{16} + \frac{7}{16} =$

5. $\frac{4}{7} + \frac{1}{3} =$

6. $\frac{3}{5} + \frac{5}{6} =$

7. $\frac{2}{15} + \frac{3}{20} =$

8. $\frac{3}{8} + \frac{5}{6} =$

9. $\frac{4}{21} + \frac{1}{6} =$

10. $\frac{11}{33} + \frac{1}{3} + \frac{9}{11} =$

Name ________________ **Date** ________ **Class** ________

11. $\frac{37}{96} + \frac{11}{16} + \frac{7}{8} =$

12. $\frac{15}{32} + \frac{7}{8} + \frac{3}{4} =$

13. $7\frac{2}{5} + 12\frac{5}{6} =$

14. $18\frac{9}{16} + 19\frac{55}{112} =$

15. $27\frac{11}{32} + 56\frac{3}{4} + 78\frac{3}{16} =$

16. $8\frac{13}{15} + 23\frac{7}{9} + 41\frac{2}{3} =$

17. $\frac{3}{4}'' + \frac{5}{16}'' =$

18. $3\frac{1}{3}$ hours + $2\frac{1}{2}$ hours + $4\frac{1}{4}$ hours =

19. $6\frac{3}{8}'' + 5\frac{9}{16}'' =$

20. $27\frac{1}{3}$ bd ft + $18\frac{5}{9}$ bd ft =

Name ______________________ **Date** ____________ **Class** ____________

21. $19\frac{3}{4}'' + 27\frac{5}{8}'' + 13\frac{13}{16}'' =$

22. What is the thickness of a floor consisting of 3/4″ plywood, 5/8″ concrete board, 3/16″ adhesive, and 3/8″ tile?

____

23. If two 4 5/8″ pieces are cut from a board, allowing for 1/8″ saw kerfs, how much of the board has been removed?

____

24. If the dimension between the finished ***jambs**** in the image shown is 28 1/2″, what is the total length of the window ***stool****, taking into account the 1/4″ setback, 2 1/4″ casing, and 1 1/8″ returns?

***See Appendix A**
Interior Finish

____

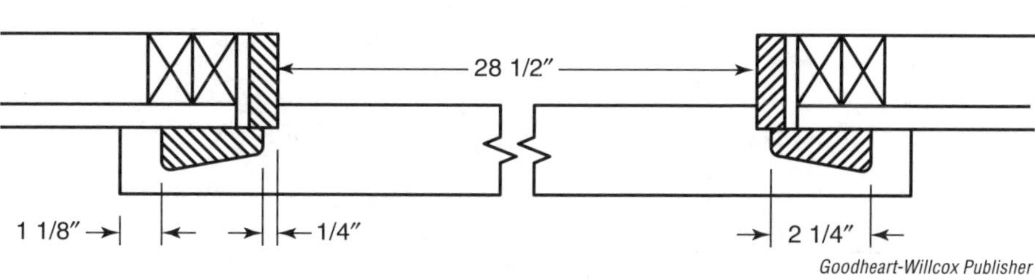

Goodheart-Willcox Publisher

25. A contractor is estimating concrete for a project that requires 10 2/3 cu yd for a garage slab, 2 1/3 cu yd for a sidewalk, and 2 2/3 cu yd for a patio. How many total cubic yards of concrete are needed for the project?

____

Use the following partial foundation plan to answer problems 26–29.

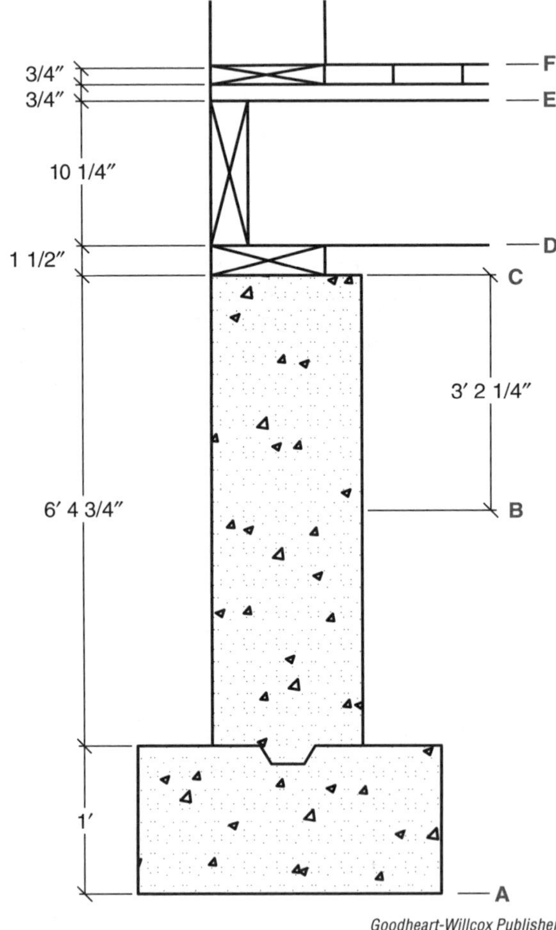

Goodheart-Willcox Publisher

26. What is the measurement from the bottom of the footer (A) to the bottom of the floor joist (D)?

____

Name ______________________ Date ______________ Class ______________

27. What is the measurement from the crawl space floor (B) to the bottom of the floor joist (D)?

____

28. What is the measurement from the top of the foundation wall (C) to the top of the floor joist (E)?

____

29. What is the measurement from the bottom of the floor joist (D) to the top of the finished floor (F)?

____

30. What is the height of a wall with 92 5/8″ studs and three 1 1/2″ plates?

____

31. What is the thickness of the exterior wall in the following figure?

____

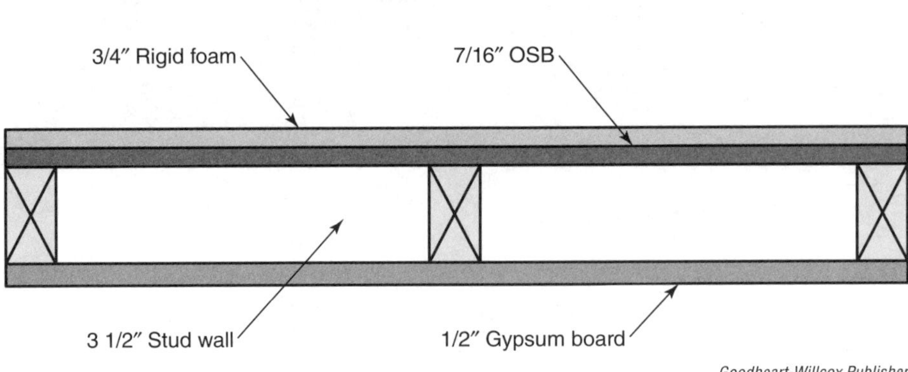

Goodheart-Willcox Publisher

***See Appendix A**
Foundation and Floor Frame

32. In the following figure, how wide is the built up ***beam**** that consists of three 2 × 10 boards and two layers of 3/4″ plywood? (Note: 2 × 10 boards are actually 1 1/2″ wide.)

____

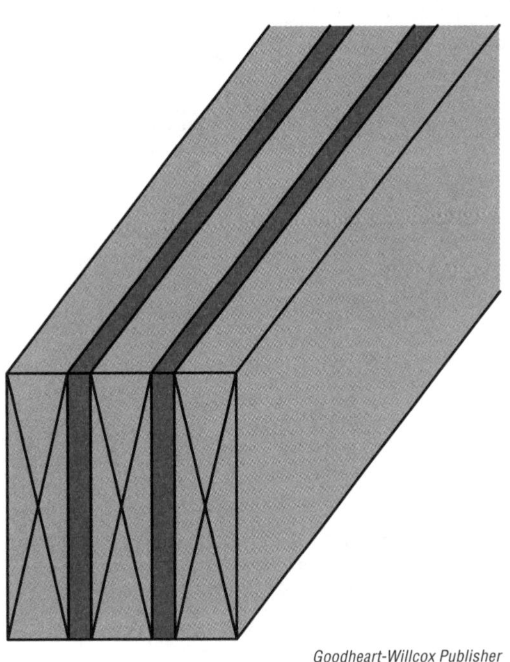

Goodheart-Willcox Publisher

Name ______________________ Date ____________ Class ____________

33. How wide is the ***rough opening**** for a 2′6″ door allowing for 3/4″ jambs, and 1/2″ clearance on each side?

*See Appendix A
Wall Frame

____

Use the following plan of a staircase to answer problems 34–35.

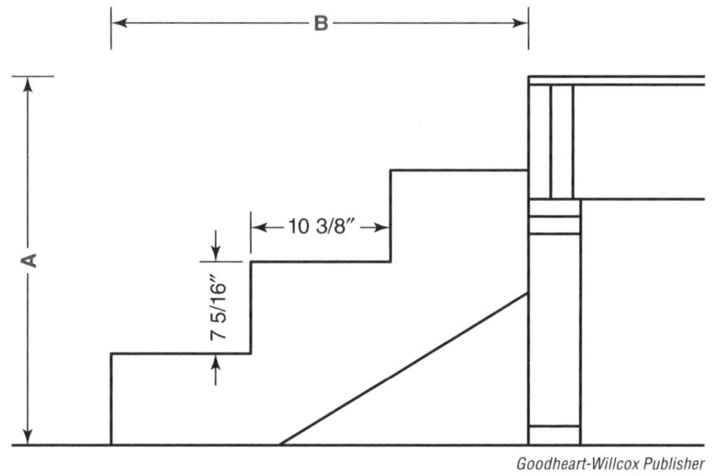

Goodheart-Willcox Publisher

34. What is the ***total rise**** of A, assuming the stairs have a uniform height?

*See Appendix A
Stair Frame

____

35. What is the ***total run**** of B, assuming the stairs have a uniform length?

*See Appendix A
Stair Frame

____

Work Space/Notes

UNIT 8

Subtracting Fractions

Objectives

After studying this unit, you will be able to:

- Demonstrate the method used to subtract fractions.
- Subtract fractions with common denominators.
- Subtract fractions without common denominators.
- Subtract mixed numbers.
- Demonstrate how to borrow from a whole number to subtract fractions.

The process of subtracting fractions involves many of the same concepts as adding fractions. A carpenter must be able to subtract fractions quickly and accurately to solve typical problems encountered on the jobsite. For example, given the total height of the wall shown in the following image, subtraction can be used to determine the stud length. The height of the three plates is subtracted from the total wall height to determine the stud length.

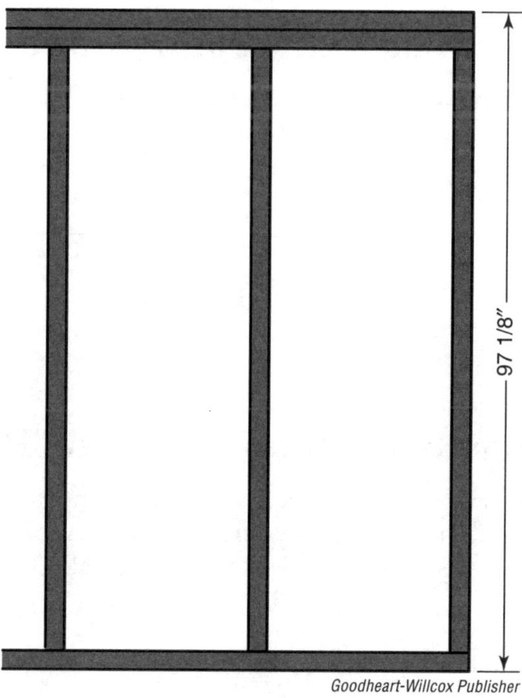

Goodheart-Willcox Publisher

Wall height	$97\frac{1}{8}''$
Thickness of plates	$-\ 4\frac{1}{2}''$
Stud length	$92\frac{5}{8}''$

Method Used to Subtract Fractions

When subtracting fractions, you must determine what type of fractions are being used. Do the fractions have common denominators (5/16 and 3/16), are the denominators different (7/8 and 5/16), or do they contain mixed numbers without a common denominator (7 11/16 and 2 3/8)? Each of these scenarios requires the use of a different process to solve the problem, but all will lead to solving a subtraction problem with common denominators.

Fractions can be subtracted by setting them up horizontally or vertically.

Horizontal subtraction:

$$\frac{7}{8} - \frac{3}{8} = ?$$

Vertical subtraction:

$$\begin{array}{r} 9\frac{13}{16} \\ -\ 4\frac{3}{8} \\ \hline ? \end{array}$$

Math Tip

A general rule for subtracting fractions is that before any subtraction can begin, all the fractions must have common denominators.

Subtracting Fractions with Common Denominators

When subtracting fractions with common denominators, the numerators are subtracted and their difference is placed over the original denominator.

Example 8-1

$$\frac{9}{16} - \frac{6}{16} = \frac{9-6}{16} = \frac{3}{16}$$

Subtracting Fractions without Common Denominators

Fractions that do not have the same denominator cannot be subtracted until the fractions are converted to have common denominators. Again, the goal is to choose a number that is the lowest common denominator (LCD) for each of the original denominators.

Example 8-2

The following fractions do not have common denominators so the LCD must be determined.

$$\frac{15}{16} - \frac{5}{6} =$$

List the multiples of each denominator to determine the LCD:

$16 \times 1 = 16$	$6 \times 1 = 6$
$16 \times 2 = 32$	$6 \times 2 = 12$
$16 \times 3 = \textcircled{48}$	$6 \times 3 = 18$
$16 \times 4 = 64$	$6 \times 4 = 24$
$16 \times 5 = 80$	$6 \times 5 = 30$
	$6 \times 6 = 36$
	$6 \times 7 = 42$
	$6 \times 8 = \textcircled{48}$
$\frac{15}{16} \times \frac{3}{3} = \frac{45}{48}$	$\frac{5}{6} \times \frac{8}{8} = \frac{40}{48}$

Once the LCD is found, each fraction is converted to an equivalent fraction using the LCD. Subtract the fractions and reduce to lowest terms if needed.

$$\frac{45}{48} - \frac{40}{48} = \frac{5}{48}$$

Subtracting Mixed Numbers

Mixed numbers whose fractions are common can be subtracted by determining the difference of the whole numbers and the fractions separately.

Example 8-3

$$\begin{array}{r} 22\frac{7}{8} \\ -\ 12\frac{5}{8} \\ \hline ? \end{array}$$

Mixed numbers can be subtracted easily by thinking of the problem as two sets of numbers to subtract.

$$\begin{array}{r} 22 \\ -\ 12 \\ \hline 10 \end{array} \qquad \begin{array}{r} \frac{7}{8} \\ -\frac{5}{8} \\ \hline \frac{2}{8} \end{array}$$

Subtract the numbers separately, then combine them to determine the final answer.

$$10 + \frac{2}{8} = 10\frac{2}{8} = 10\frac{1}{4}$$

Example 8-4

$$\begin{array}{r} 15\frac{5}{8} \\ -\ 7\frac{3}{7} \\ \hline ? \end{array}$$

Mixed numbers whose fractions do not have common denominators cannot be solved until the LCD is determined.

$8 \times 1 = 8$	$7 \times 1 = 7$
$8 \times 2 = 16$	$7 \times 2 = 14$
$8 \times 3 = 24$	$7 \times 3 = 21$
$8 \times 4 = 32$	$7 \times 4 = 28$
$8 \times 5 = 40$	$7 \times 5 = 35$
$8 \times 6 = 48$	$7 \times 6 = 42$
$8 \times 7 = \textcircled{56}$	$7 \times 7 = 49$
$8 \times 8 = 64$	$7 \times 8 = \textcircled{56}$
$\frac{5}{8} \times \frac{7}{7} = \frac{35}{56}$	$\frac{3}{7} \times \frac{8}{8} = \frac{24}{56}$

$$\begin{array}{rcr} 15\frac{5}{8} & = & 15\frac{35}{56} \\ -\ 7\frac{3}{7} & = & -\ 7\frac{24}{56} \\ \hline & & 8\frac{11}{56} \end{array}$$

Borrowing from Whole Numbers to Subtract Fractions

Complications occur in subtracting mixed numbers when the fraction being subtracted is greater than the one it is being subtracted from. In these situations, borrowing must be used.

Borrowing in mixed numbers is similar to borrowing in whole numbers: you "borrow" from the ones place and add one to the fraction. When the whole unit is carried over to the fraction, it must be expressed as a fraction equivalent to 1, or 1/1.

Example 8-5

$$\begin{array}{r} 13\frac{3}{8} \\ -\ 8\frac{7}{8} \\ \hline ? \end{array}$$

3/8 is less than 7/8, so borrowing must occur.

$$\begin{array}{r} \overset{12}{\cancel{13}}\frac{3}{8}^{+\frac{8}{8}} \\ -\ 8\frac{7}{8} \\ \hline ? \end{array}$$

After borrowing and converting the 3/8 to 11/8, the subtraction problem becomes easy to complete.

$$\begin{array}{r} 12\frac{11}{8} \\ -\ 8\frac{7}{8} \\ \hline 4\frac{4}{8} \end{array} = 4\frac{1}{2}$$

Work Space/Notes

Unit 8 Review

Name ____________________ Date __________ Class __________

Solve the following equations and reduce the answer to lowest terms. Show all of your work.

1. $\frac{3}{4} - \frac{1}{4} =$

2. $\frac{13}{16} - \frac{7}{16} =$

3. $\frac{33}{64} - \frac{21}{64} =$

4. $\frac{17}{25} - \frac{12}{25} =$

5. $\frac{11}{16} - \frac{3}{8} =$

6. $\frac{15}{18} - \frac{41}{72} =$

7. $\frac{5}{8} - \frac{13}{64} =$

8. $\frac{7}{12} - \frac{5}{36} =$

9. $14\frac{5}{9} - 7\frac{3}{9} =$

10. $23\frac{3}{7} - 19\frac{1}{7} =$

Name ______________________ **Date** ____________ **Class** ____________

11. $123\frac{5}{12} - 96\frac{7}{12} =$

12. $48\frac{3}{16} - 27\frac{9}{16} =$

13. $36\frac{1}{4} - 15\frac{9}{32} =$

14. $15\frac{5}{16} - 2\frac{3}{4} =$

15. $13\frac{5}{8}'' - 8\frac{1}{4}'' =$

16. $17\frac{3''}{16} - 4\frac{7''}{8} =$

17. $63\frac{1}{4}$ sq ft – $18\frac{2}{3}$ sq ft =

18. $27\frac{3}{4}$ bd ft – $11\frac{15}{16}$ bd ft =

19. A board measuring 1 3/16″ thick was run through a surface planer three times removing 1/8″ on each pass through the planer. What is the final thickness of the board?

____

20. Three cuts are made from an 18″ board. Allowing for 1/8″ saw kerfs, how much of the board is left after cutting the 3 1/4″, 6 1/2″, and 4 7/8″ pieces?

____

Name ______________________ **Date** ____________ **Class** ____________

21. If a header is installed in a wall 82 1/4″ off the floor, how long is the ***trimmer (jack stud)**** supporting the header allowing for a 1 1/2″ sole plate?

***See Appendix A**
Wall Frame

____

Use the following dimensioned block to answer questions 22–25.

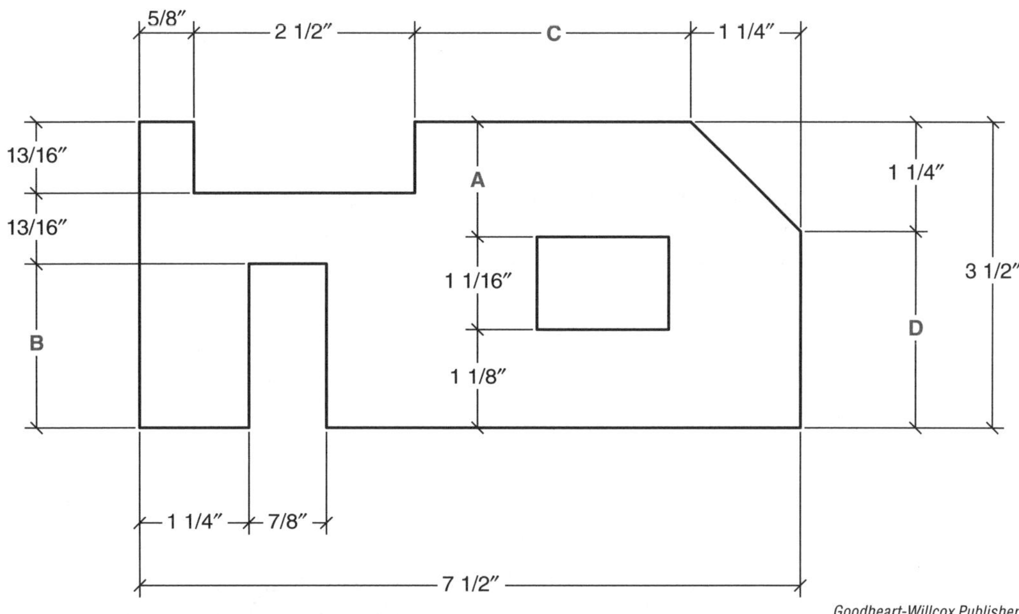

Goodheart-Willcox Publisher

22. What is the dimension of Distance A?

____

23. What is the dimension of Distance B?

____

24. What is the dimension of Distance C?

____

25. What is the dimension of Distance D?

____

26. A house requires 48 squares of shingles to cover the roof. The roofing crew installs 17 1/3 squares on Monday and 22 1/2 squares on Tuesday. How many squares need to be installed on Wednesday to complete the job?

____

Name ______________________ **Date** ____________ **Class** ____________

27. Using a jointer, a 9 1/4″ board has 3/8″ removed from one edge and 5/16″ off the other edge. What is the remaining width of the board?

____

Use the dimensions and floor system shown in the following plan to answer questions 28–29.

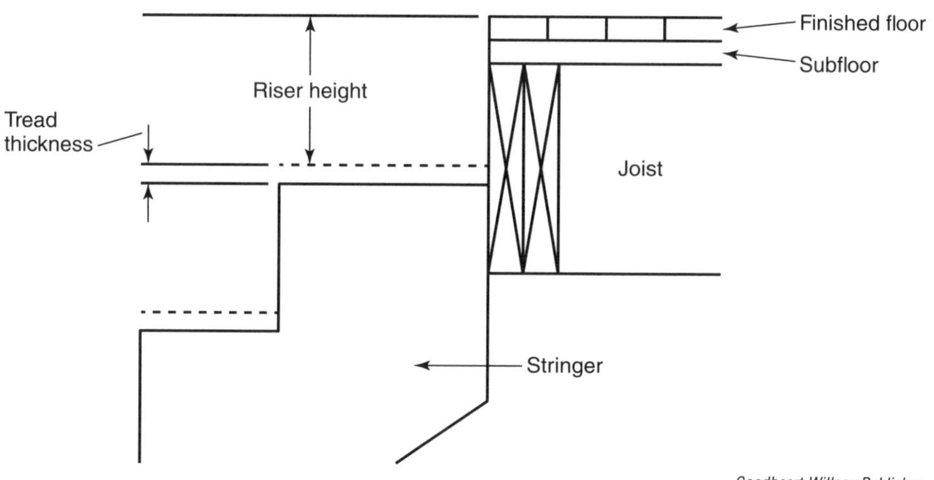

Goodheart-Willcox Publisher

Riser* height—7 1/4″

Finished floor thickness—3/4″

Subfloor thickness—11/32″

Tread* thickness—7/8″

*See Appendix A
Stair Frame

*See Appendix A
Stair Frame

28. How far down from the top of the subfloor does the ***stringer**** need to be attached to maintain a 7 1/4″ riser height?

*See Appendix A
Stair Frame

____

29. If the overall floor system measures 12 11/32″, what is the thickness of joist?

____

30. A finished exterior wall measures 5 3/16″ thick. The studs measure 3 1/2″, the drywall is 1/2″, and the exterior-rated sheathing is 7/16″. How thick is the layer of rigid foam applied to the outside of the sheathing?

____

UNIT 9

Multiplying Fractions

Objectives

After studying this unit, you will be able to:

- Demonstrate the method used to multiply fractions.
- Multiply mixed numbers.
- Multiply more than two fractions.
- Multiply more than two mixed numbers.

The process of multiplying fractions is much easier than adding or subtracting, because common denominators are not required. To multiply two fractions, simply multiply the numerators to calculate the numerator of the product and then multiply the denominators to calculate the denominator of the product. The product of this operation is reduced to its lowest terms. Before mixed numbers can be multiplied, they must first be converted to improper fractions.

For example, the height of a set of stairs can be determined using multiplication. The vertical distance from one floor level to another, ***total rise****, is calculated by multiplying the vertical distance of one riser, ***unit rise****, by the total number of risers.

*See Appendix A
Stair Frame

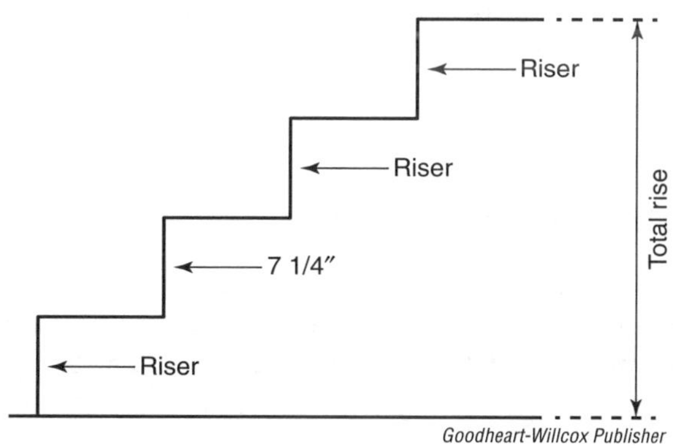

Goodheart-Willcox Publisher

7 1/4″ riser × 4 = Total Rise

Method Used to Multiply Fractions

When multiplying fractions, the fractions are typically arranged horizontally, multiplied, and reduced when needed.

Example 9-1

$$\frac{7}{8} \times \frac{3}{4} = \frac{21}{32}$$

Once a multiplication problem is set up, the ability to **cancel** numbers in a fraction can reduce the size of the numbers to be multiplied. To help see this, combine the two fractions as a single fraction.

Example 9-2

$$\frac{3}{16} \times \frac{32}{33} = \frac{3 \times 32}{16 \times 33}$$

Before multiplying the fraction, look for any opportunities to cancel numbers that will divide evenly into the numerator and the denominator. After each reduction is made, look for other opportunities to simplify the problem until the fraction has been completely reduced. Once completely simplified, perform the multiplication.

$$\frac{\overset{1}{\cancel{3}} \times 32}{16 \times \underset{11}{\cancel{33}}} = \frac{1 \times \overset{2}{\cancel{32}}}{\underset{1}{\cancel{16}} \times 11} = \frac{1 \times 2}{1 \times 11}$$

Taking the time to reduce the problem to its lowest possible terms makes the multiplication process much easier.

$$\frac{1 \times 2}{1 \times 11} = \frac{2}{11}$$

Math Tip

Always look for opportunities to cancel numbers in fractions before multiplying, then reduce the product if needed.

Multiplying Mixed Numbers

When multiplying mixed numbers, they must first be converted to improper fractions. Be sure to check the improper fractions for any opportunities to reduce by canceling. Once the multiplication is complete, convert the product back to a mixed number and reduce if needed.

Example 9-3

$$3\frac{5}{7} \times 2\frac{3}{4} = \frac{26}{7} \times \frac{11}{4}$$

$$= \frac{\overset{13}{\cancel{26}} \times 11}{7 \times \underset{2}{\cancel{4}}}$$

$$= \frac{13 \times 11}{7 \times 2}$$

$$= \frac{143}{14}$$

$$= 10\frac{3}{14}$$

Multiplying More than Two Fractions

The process for multiplying more than two fractions follows the same pattern described for two-fraction multiplication. The fractions are combined into a single fraction, numbers are canceled if possible, the numerators and denominators are multiplied, and the product is reduced if needed.

Example 9-4

$$\frac{7}{16} \times \frac{6}{7} \times \frac{5}{8} = \frac{7 \times 6 \times 5}{16 \times 7 \times 8}$$

$$= \frac{\overset{1}{\cancel{7}} \times 6 \times 5}{16 \times \underset{1}{\cancel{7}} \times 8}$$

$$= \frac{1 \times \overset{3}{\cancel{6}} \times 5}{\underset{8}{\cancel{16}} \times 1 \times 8}$$

$$= \frac{1 \times 3 \times 5}{8 \times 1 \times 8}$$

$$= \frac{15}{64}$$

Multiplying More than Two Mixed Numbers

When multiplying more than two mixed numbers, the mixed numbers must first be converted to improper fractions and combined within one fraction.

Example 9-5

$$3\frac{1}{16} \times 4\frac{4}{7} \times 3\frac{3}{4} = \frac{49}{16} \times \frac{32}{7} \times \frac{15}{4}$$

$$= \frac{\overset{7}{\cancel{49}} \times 32 \times 15}{16 \times \underset{1}{\cancel{7}} \times 4}$$

$$= \frac{7 \times \overset{2}{\cancel{32}} \times 15}{\underset{1}{\cancel{16}} \times 1 \times 4}$$

$$= \frac{7 \times 2 \times 15}{1 \times 1 \times 4}$$

$$= \frac{210}{4}$$

$$= 52\frac{1}{2}$$

Unit 9 Review

Name ____________________ Date ____________ Class ____________

Solve the following equations and reduce the answer to lowest terms. Show all of your work.

1. $\frac{3}{8} \times \frac{1}{2} =$

2. $\frac{4}{13} \times \frac{5}{16} =$

3. $\frac{12}{25} \times \frac{5}{32} =$

4. $\frac{7}{8} \times \frac{15}{28} =$

5. $\frac{1}{32} \times \frac{12}{19} =$

6. $\frac{7}{12} \times \frac{5}{8} =$

7. $6\frac{3}{7} \times 5\frac{5}{8} =$

8. $3\frac{3}{16} \times 1\frac{7}{12} =$

9. $17\frac{1}{4} \times 12\frac{6}{9} =$

10. $4\frac{11}{12} \times 2\frac{3}{7} =$

Name ______________________________ **Date** ______________ **Class** ______________

11. $9\frac{3}{8} \times 5\frac{15}{16} =$

12. $7 \times 18\frac{1}{8} =$

13. $6\frac{1}{4} \times 3\frac{3}{8} \times 5\frac{3}{5} =$

14. $2\frac{4}{5} \times 3\frac{5}{6} \times 5\frac{7}{8} =$

15. $\frac{15}{16} \times 6 \times 4\frac{1}{4} =$

16. $6\frac{3}{8} \times \frac{7}{8} \times 17 =$

17. $2\frac{5}{8} \times 16\frac{1}{4} \times 8\frac{3}{4} =$

18. $4\frac{1}{3} \times \frac{9}{16} \times 2\frac{3}{5} =$

19. A wall is finished with 19 horizontal panels, each measuring 5 1/2″. What is the height from the floor to the upper edge of the top panel?

____

Name ______________________ **Date** ____________ **Class** ____________

Use the following plan of a staircase to answer questions 20–21.

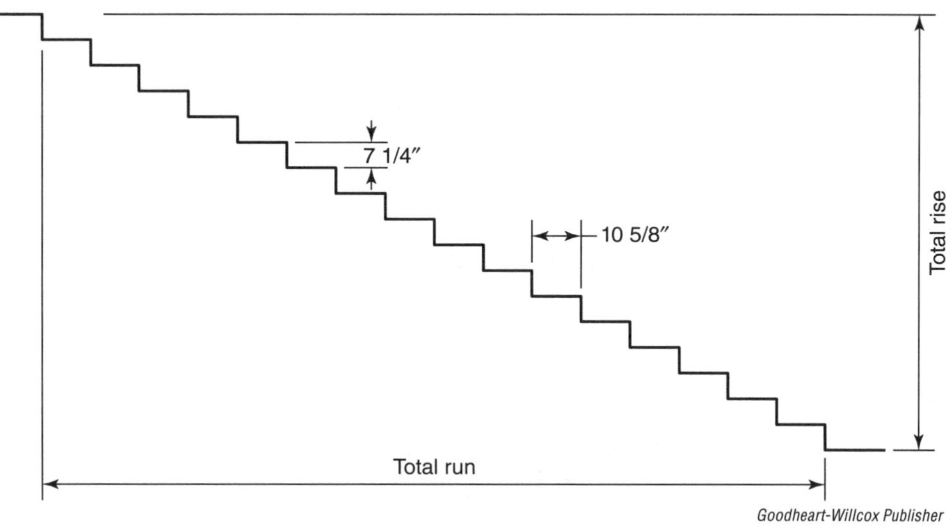

Goodheart-Willcox Publisher

20. Using the measurements shown, what is the total rise from floor to floor?

____

21. Using the measurements shown, what is the total run of the stairs?

____

22. It takes 27 rows of shingles to cover one side of a ***gable roof**** to the ***ridge****. With the shingles spaced at a 5 1/2″ exposure, what is the measurement from the ***drip edge**** along the eave to the ridge?

____

***See Appendix A**

Roofing Types and Terminology; Roof Frame; and Exterior Finish.

23. A foyer has 18 rows of 8″ square ceramic tile installed with 3/8″ grouted seams. How wide is the foyer?

____

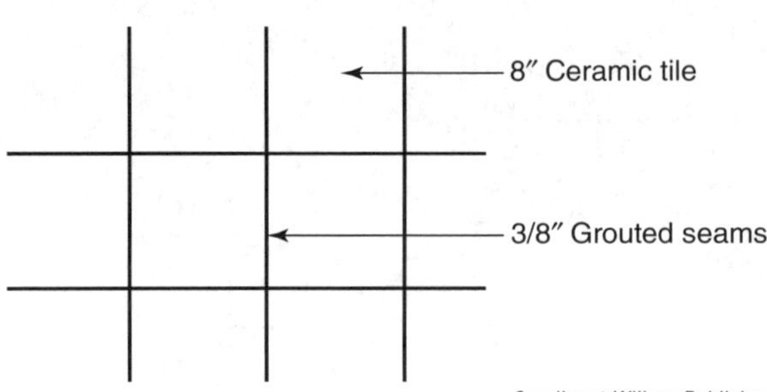

Goodheart-Willcox Publisher

24. Six 1 × 10 shelves must be cut 18 3/4″ for a shelving unit. Allowing for 1/8″ saw kerfs, how much 1 × 10 material is needed?

____

Use the fence shown to answer questions 25–26.

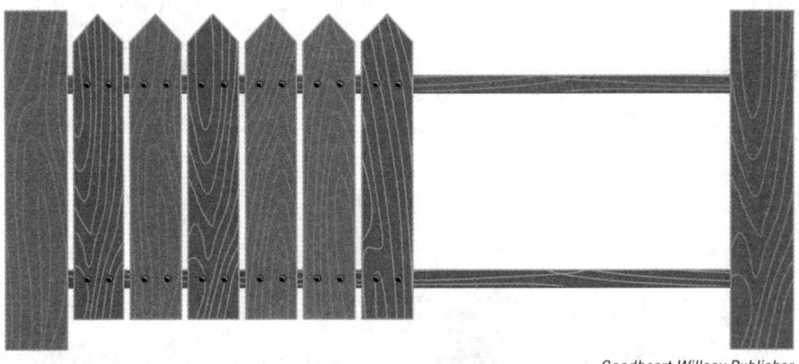
Goodheart-Willcox Publisher

25. Twelve 3 1/2″ vertical fence slats are installed between the support posts with 1/4″ spacing between each slat and between slats and the posts. How far apart are the posts?

____

Name ______________________ Date ____________ Class ____________

26. Twelve 7 1/4″ vertical fence slats are installed between the support posts with 3/8″ spacing between each slat and the posts. How far apart are the posts?

____

27. A set of porch steps consists of 6 risers each measuring 7 9/16″. How far above grade (the height of the existing ground) is the porch?

____

28. A retaining wall is to be constructed out of 5 1/2″ thick blocks and will be 9 rows high. What will the finished height of the wall be?

____

29. ***Cap**** shingles are applied to the ridge of a house with an exposure of 5 3/4″ per cap. How long is the ridge of the roof if 96 caps are used?

____

*See Appendix A
Exterior Finish

30. The following sheet goods are delivered to a wood shop: twenty-two 3/4″ OSB, twenty-four 5/8″ plywood, eighteen 1/2″ plywood. What does the stack measure if all the sheets are placed in one stack?

____

UNIT 10

Dividing Fractions

Objectives

After studying this unit, you will be able to:

- Demonstrate the method used to divide fractions.
- Solve division problems containing mixed numbers.

Once the process of multiplying fractions has been learned, dividing fractions becomes easy to learn by adding one simple step to the process.

Method Used to Divide Fractions

When performing a fractional division operation, the fractions are set up horizontally and are separated by the division symbol.

$$\frac{7}{9} \div \frac{4}{5} = ?$$

The first step in solving the equation is to find the **reciprocal** of the devisor. The reciprocal is the value to multiply a number in order to have a product of 1. To find the reciprocal, simply invert the fraction by switching the positions of the numerator and denominator. For example, the reciprocal of 4/5 is 5/4.

$$\frac{4}{5} \times \frac{5}{4} = \frac{20}{20} = 1$$

Math Tip

The reciprocal of whole numbers can be found as easily as that of fractions. Remember that a whole number can be represented as a fraction that has the whole number as the numerator and one as the denominator. For example, 4 = 4/1. By expressing a whole number as a fraction, the reciprocal becomes more clear. For example, the reciprocal of 4/1 is 1/4.

After the reciprocal has been found, the division problem becomes one of multiplication. Multiply the dividend by the reciprocal of the divisor to find the solution to the equation.

Example 10-1

$$\frac{7}{9} \div \frac{4}{5} = ?$$

$$\frac{7}{9} \div \frac{4}{5} = \frac{7}{9} \times \frac{5}{4} = \frac{35}{36}$$

As with multiplying fractions, look for opportunities to simplify the fractions after the reciprocal has been found. This can reduce the size of the numbers before multiplying.

Example 10-2

$$\frac{1}{3} \div \frac{5}{12} = \frac{1}{3} \times \frac{12}{5}$$

$$= \frac{1 \times \overset{4}{\cancel{12}}}{\underset{1}{\cancel{3}} \times 5}$$

$$= \frac{1 \times 4}{1 \times 5}$$

$$= \frac{4}{5}$$

Math Tip

When dividing fractions, invert the divisor and change the operational symbol from division to multiplication.

Dividing Mixed Numbers

As with other operations, when dividing fractions with mixed numbers, the fractions must first be converted into improper fractions. Then perform the division on the improper fractions.

Example 10-3

All mixed numbers in the operation are converted to improper fractions.

$$2\frac{5}{8} \div 3\frac{2}{3} = \frac{21}{8} \div \frac{11}{3}$$

The next step is to invert the divisor and change the operational symbol to division.

$$\frac{21}{8} \div \frac{11}{3} = \frac{21}{8} \times \frac{3}{11}$$

Finally, multiply the fractions and reduce the answer as needed.

$$\frac{21 \times 3}{8 \times 11} = \frac{63}{88}$$

Unit 10 Review

Name ______________________ Date __________ Class __________

Solve the following equations and reduce the answer to lowest terms. Show all of your work.

1. $\frac{5}{12} \div \frac{3}{4} =$

2. $\frac{4}{9} \div \frac{6}{7} =$

3. $\frac{1}{16} \div \frac{3}{32} =$

4. $\frac{12}{17} \div \frac{4}{5} =$

5. $\frac{5}{8} \div \frac{3}{5} =$

6. $\frac{7}{13} \div \frac{4}{5} =$

7. $4 \div \frac{3}{8} =$

8. $7 \div \frac{5}{12} =$

9. $16 \div 2\frac{7}{8} =$

Name ______________________ **Date** ____________ **Class** ____________

10. $23 \div 8\frac{1}{3} =$

11. $14\frac{5}{8} \div 3 =$

12. $6\frac{5}{9} \div 21 =$

13. $4\frac{3}{4} \div 2\frac{1}{8} =$

14. $1\frac{2}{5} \div 4\frac{7}{8} =$

15. $9\frac{1}{8} \div 3\frac{3}{4} =$

16. $2\frac{1}{5} \div 7\frac{2}{3} =$

17. $4\frac{5}{6} \div 12\frac{1}{6} =$

18. $5\frac{2}{3} \div 18\frac{1}{2} =$

19. How many 6 1/2″ blocks can be cut from a 47 1/2″ board?

____

Name ______________________ Date ____________ Class ____________

Use the following stairs to answer questions 20–21.

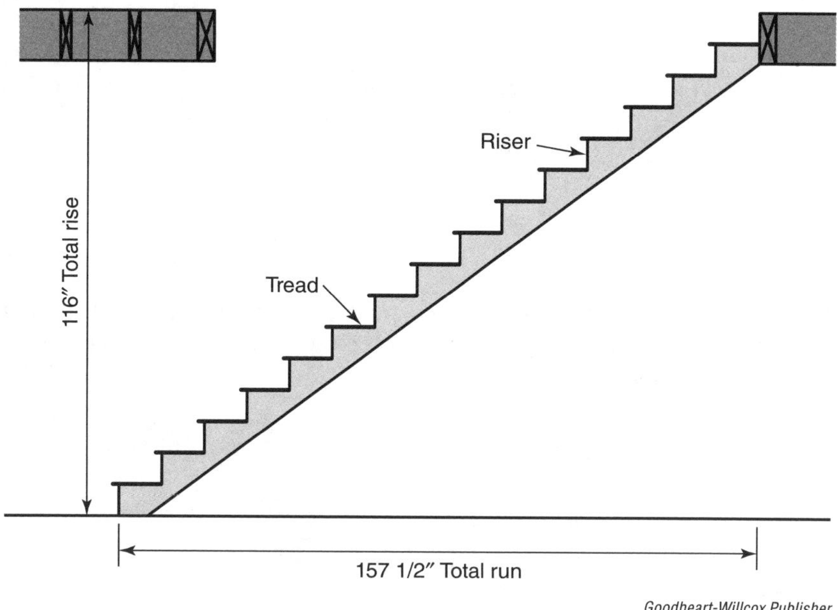

20. The stairs have a 116″ total rise. What is the riser dimension?

____

21. The stairs have a 157 1/2″ total run. What is the tread dimension?

____

22. A standard concrete block measures 15 5/8″. Using a 3/8″ mortar joint between each block, how many blocks are needed to lay one row on a 192″ foundation?

____

23. A roof that measures 187″ from the drip edge, along the eave to the ridge is shingled using a 5 1/2″ exposure on each row. How many rows are needed to cover the roof surface?

____

24. How many risers are in a set of stairs if the total rise is 92 5/8″ and the risers are 7 1/8″?

____

Name ______________________ Date ______________ Class ______________

25. In the following image of a railing system, each baluster measures 1 1/2″ wide. Assuming equal spacing, what is the distance between each post?

____

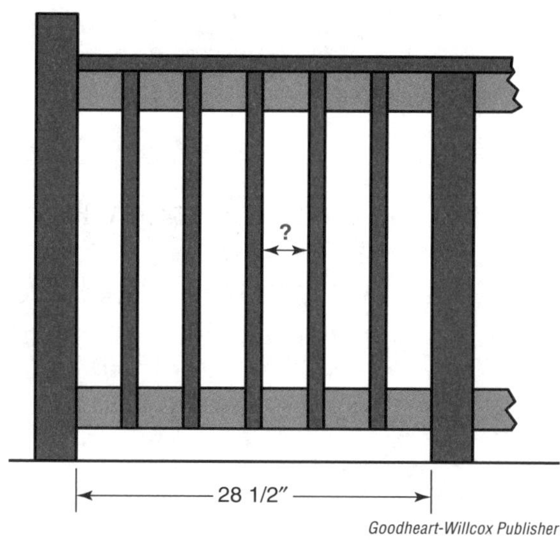

Goodheart-Willcox Publisher

26. How many treads are in a set of stairs if the total run is 124 1/2″ and the treads are each 10 3/8″?

____

27. How many rows of 2 1/4″ boards are needed to cover a floor that measures 78 3/4″ wide?

____

28. A ridge measuring 126″ long will be capped using a 5 1/4″ exposure on each cap. How many caps are needed to cap the ridge?

____

29. A carpenter worked 17 3/4 hours and grossed $426. What was his hourly wage?

____

30. A board measuring 116 7/8″ needs to be cut into 8 equal pieces. Allowing for 1/8″ saw kerfs, how long will the pieces be?

____

Section 2 Exam

Name ______________________ Date ____________ Class ____________

1. Find the missing numerator.

   $\frac{5}{8} = \frac{\quad}{72}$

2. Reduce 4/64 to lowest terms.

   ____

3. Convert 38/7 to a mixed number.

   ____

4. Convert 7 3/8 to an improper fraction.

   ____

5. Determine the lowest common denominator (LCD) for the following fractions.

   $\frac{7}{8}$ $\quad$ $\frac{5}{12}$ $\quad$ $\frac{2}{9}$

6. $\frac{7}{18} + \frac{5}{18} + \frac{11}{18} =$

7. $8\frac{5}{8} + 22\frac{3}{4} =$

8. $9\frac{1}{4}' + 12\frac{5}{12}' =$

9. $\frac{13}{32} - \frac{5}{16} =$

10. $114\frac{7}{8} - 66\frac{5}{16} =$

Name ______________________ **Date** ____________ **Class** ____________

11. $28\frac{5}{16}'' - 12\frac{5}{8}'' =$

12. $\frac{7}{16} \times \frac{3}{8} =$

13. $6\frac{3}{4} \times 4\frac{8}{9} =$

14. $5\frac{3}{5} \times \frac{15}{16} \times 6\frac{2}{5} =$

15. $\frac{23}{32} \div \frac{3}{16} =$

16. $62 \div 7\frac{3}{4} =$

17. $8\frac{3}{4} \div 2\frac{1}{3} =$

18. A general contractor needs to order concrete for multiple projects this week. The first requires 14 1/3 cu yd, another requires 22 2/3 cu yd, and a third requires 18 2/3 cu yd. How many total cubic yards of concrete does the general contractor need to order?

____

19. What is the height of a wall with 104 3/8″ studs and three 1 1/2″ plates?

____

Name ______________________ Date ____________ Class ____________

20. A roofing crew installed 14 2/3 squares of shingles on Tuesday and 17 1/2 squares on Wednesday. To complete the job, they will need to install a total of 60 squares. How many squares still need to be installed to the complete the job?

____

21. Using a surface planer, a 3 1/8″ thick board has 5/16″ removed from one surface and 3/16″ off the other surface. What is the final thickness of the planed board?

____

22. A set of basement steps with a landing consists of 7 risers from the cellar floor to the top of the landing. If the risers each measure 7 3/8″, what is the height of the landing?

____

23. Architectural shingles with an exposure of 5 1/2″ were used to shingle a roof. What is the measurement from the drip edge at the eve to the ridge if 23 rows of shingles were used?

____

24. A roofer worked 39 1/2 hours for the week and grossed $711 on his paycheck. What is his hourly wage?

____

25. A carpenter is installing 4 1/4″ T & G hardwood flooring in an office that measures 172 1/2″ wide. How many rows of flooring are needed to cover the subfloor?

____

SECTION 3

Decimals

Unit 11. Basic Principles of Decimals
Unit 12. Adding and Subtracting Decimals
Unit 13. Multiplying Decimals
Unit 14. Dividing Decimals

Key Terms

decimal
decimal point

UNIT 11

Basic Principles of Decimals

Objectives

After studying this unit, you will be able to:

- Determine place value of decimals.
- Demonstrate how to round decimals to a certain place value.

Our numbering system consists of whole numbers, common fractions, and mixed numbers. This unit introduces another classification of numbers called decimal numbers.

In our numbering system, the value of a digit is determined by its location or place value in the number. The decimal numbering system uses the same method of place values to identify values less than one. These numbers are identified as a decimal fraction, but are commonly referred to as simply a **decimal**.

All decimal numbers include a **decimal point**. Digits to the left of the decimal point are in place values equal to or greater than one and digits to the right of the decimal point (decimals) are in place values less than one.

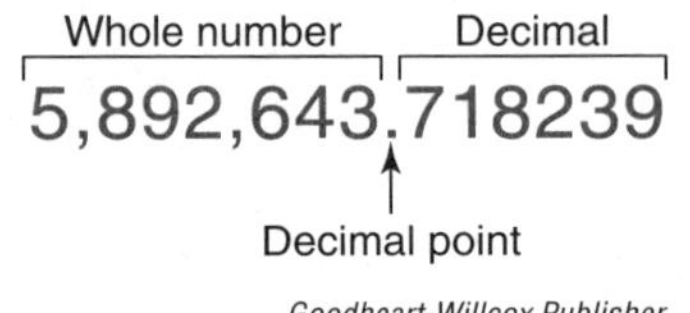

Goodheart-Willcox Publisher

The value of a decimal can be verbally expressed in two ways:

- The digits can be pronounced individually as they appear in the number. For example, 23.74 is read as "twenty-three point seven four."
- The digits can be pronounced as the whole number portion, followed by "and," followed by the decimal portion with the place value of the last decimal number. For example, 23.74 is read as "twenty-three and seventy-four hundredths."

Place Value and Rounding Decimals

Like whole numbers, decimals are identified by their location or place value. As shown in the following place value chart, each digit to the right of the decimal has a value and a name based on units of ten.

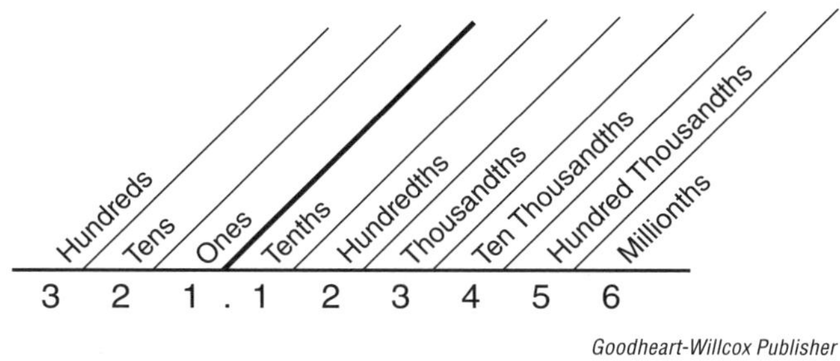

Goodheart-Willcox Publisher

Math Tip

A decimal point can also be referred to as a “point” or by using the word “and.”

Often when performing mathematic calculations, numbers have decimals that become quite long, extending to the ten thousandths place value or farther. When this occurs, it is common practice to round the decimal to a certain degree of accuracy. This degree of accuracy is determined by the type of work being performed on the jobsite.

The process to round a decimal is:

1. Determine the place value that will be rounded to.
2. Look one digit to the right of that place value. If that number is 5 or greater, add one to the final number in your answer. If the number to the right is 4 or less, simply round to the desired place value.

Example 11-1

425.94835

Round the number to the hundredths place value. Initially, by cutting off the number at the hundredths place value, the answer is 425.94~~835~~.

Since the number one digit to the right of the hundredths column is greater than 5, the 4 is increased by one to 5.

The final answer is 425.95.

Example 11-2

168.338442

Round the decimal to the nearest thousandth. Initially, by cutting off the number at the thousandths place value, the answer is 168.338~~442~~.

Since the number one digit to the right of the thousandths column is less than 5, the 8 does not change. Rounding to the thousandths produces the answer 168.338.

Converting Decimals to Fractions

There are times when a decimal must be converted to a fraction. The decimal is placed into a fraction as the numerator and the last place value is inserted as the denominator.

$$.3 = \frac{3}{10}$$

$$.03 = \frac{3}{100}$$

$$.003 = \frac{3}{1000}$$

Because place values are based on units of 10, the denominators will follow the place value sequence of 10, 100, 1,000, 10,000, and so on. Once the decimal fraction has been converted, reduce the common fraction to its lowest terms as needed.

Example 11-3

$$.6 = \frac{6}{10} = \frac{3}{5}$$

$$.42 = \frac{42}{100} = \frac{21}{50}$$

$$.144 = \frac{144}{1000} = \frac{18}{125}$$

Carpentry Notes

Calculating Rafter Length

***See Appendix A**

Roof Types and Terminology; Roof Frame

In the process of calculating the ***line length**** for a ***common rafter****, the variables are multiplied in decimal form. The line length is then converted into a fraction so the length can be found on a tape measure.

To determine the line length of a common rafter, multiply half of the *span* (***total run****) by the ***unit length****.

Building span is 25′ 0″

Total run is 12.5′

Unit length off framing square for an 8/12 pitch roof is 14.42″

$$12.5 \times 14.42 = 180.25''$$

$$180.25 = \frac{25}{100} = \frac{1}{4} = 180\frac{1}{4}''$$

Unit 11 Review

Name ______________________ Date ____________ Class ____________

Write the following as decimal numbers.

1. Six point seven three nine.

 ____

2. Two hundred point zero one zero nine.

 ____

3. Zero point six six two eight four.

 ____

4. Eleven and one hundred twenty-three thousandths.

 ____

5. Two thousand three hundred fourteen ten thousandths.

 ____

Express the following decimals as common fractions.

6. 0.75

 ____

7. 0.365

 ____

8. 0.1244

____

9. 0.768

____

10. 0.354

____

11. 2.9375

____

Name ______________________ **Date** __________ **Class** __________

Round the following decimals to the nearest tenth.

12. 0.0631

____

13. 0.26195

____

14. 0.55611

____

Round the following decimals to the nearest hundredth.

15. 0.02369

____

16. 0.945217

____

17. 0.941624

____

Round the following decimals to the nearest thousandth.

18. 0.5469725

____

19. 0.1973482

____

20. 0.2169056

____

21. 0.8632951

____

Solve the following problems.

22. A board must be measured at 27.625″. What is this measurement in fraction form?

____

23. The line length for a common rafter measures 114.875". What fractional measurement would the carpenter mark on the tape measure?

____

*See Appendix A
Stair Frame

24. The ***unit rise**** on a set of stairs is 7.375". What is the unit rise in fraction form?

____

25. A blueprint calls for a 0.3125" expansion gap on a concrete joint. What is the size of the gap in fractional form?

____

UNIT 12

Adding and Subtracting Decimals

Objectives

After studying this unit, you will be able to:

- Demonstrate the method used to add decimals.
- Demonstrate the method used to subtract decimals.

The process of adding and subtracting numbers with decimals uses the same concepts that are used for adding and subtracting whole numbers. When adding numbers, carrying is used to move the second digit of a sum over a column to the left. When subtracting numbers, borrowing is used to take one unit from the place value to the left of the column you are subtracting. The key difference in adding or subtracting numbers with decimals is that the decimals must be lined up vertically.

Method Used to Add Decimals

As with adding whole numbers, decimals can be added horizontally or vertically. Vertical addition aligns the decimals allowing for simple addition. Horizontal addition will generally require a calculator to solve the problem and keep the decimal in the correct place.

Horizontal addition:

$$25.3 + 0.094 + 537.324 = 562.718$$

Vertical addition:

$$\begin{array}{r} 25.300 \\ 0.094 \\ \underline{+\ 537.324} \\ 562.718 \end{array}$$

When adding decimal numbers, arrange the numbers vertically in a column so that the decimals are lined up. This ensures that digits are in the correct place value. Because this does not align all the numbers to the right as in adding whole numbers, zeros can be added on the right side of each number as needed. The digits are then added starting on the right column and carrying as needed. The decimal is placed in the answer in-line with the other decimals.

Incorrect	Correct
12.368	12.3680
576.3	576.3000
0.5891	0.5891
+ 201.58	+ 201.5800
?	790.8371

Method Used to Subtract Decimals

The same steps used to add decimal numbers will be used when subtracting. The numbers must first be set up vertically, ensuring the decimals are in-line. After performing the subtraction, place a decimal in the answer in-line with the other decimals.

Math Tip

Always keep decimals in-line and add zeros to fill in the right side of the problem.

Example 12-1

15.86 – 6.7142 =

Subtraction problems become easier to solve when lined up vertically.

$$\begin{array}{r} \scriptstyle 5\,9\,1 \\ 15.8600 \\ -\ \ 6.7142 \\ \hline 9.1458 \end{array}$$

Unit 12 Review

Name ______________________ Date __________ Class __________

Solve the following equations. Make sure the decimals are in-line and show all of your work.

1. $\begin{array}{r} 5.30 \\ 0.43 \\ 4.61 \\ +\ 2.24 \\ \hline \end{array}$

2. $\begin{array}{r} 6.27 \\ 4.11 \\ 7.01 \\ +\ 5.74 \\ \hline \end{array}$

3. $\begin{array}{r} 49.728 \\ 55.416 \\ 12.548 \\ +\ 62.812 \\ \hline \end{array}$

4. $\begin{array}{r} 32.351 \\ 21.812 \\ 9.601 \\ +\ 38.225 \\ \hline \end{array}$

5. $5.7 - 0.9 =$

6. $18.6 - 5.2 =$

7. $42.16 - 27.09 =$

8. 687.671 – 198.472 =

9. 1.2 + 0.8 + 5.7 =

10. 15.44 + 9.11 + 0.86 =

11. 281.719 + 66.813 + 207.009 =

12. 62.78 – 36.84 – 9.99 =

Name ______________________ Date ____________ Class ____________

13. What is the total cost of trim for a project using the following materials?

    Baseboard: $123.47

    Casings: $98.36

    Crown molding: $278.32

    ____

14. A contractor earned $1,236.24 on one job and $2,613.78 on another. What is the total amount the contractor earned for the two jobs?

    ____

15. What is the thickness of the countertop shown in the following image?

    ____

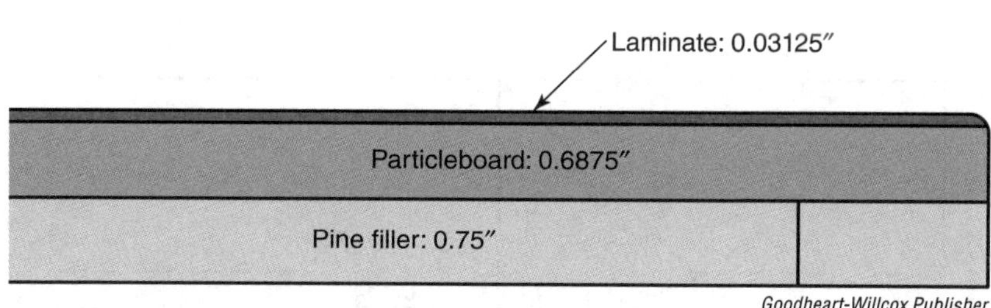

16. Hardwood flooring needs to be installed in two bedrooms measuring 175.2 sq ft and 224.9375 sq ft, and a living room measuring 263.625 sq ft. How many square feet of flooring are needed?

____

17. A contractor made two purchases for materials each totaling $714.26 and $504.98. He later returned an item for a $46.79 refund. What is the total cost for materials for this job?

____

18. What is the total square footage of the roof shown in the following figure?

____

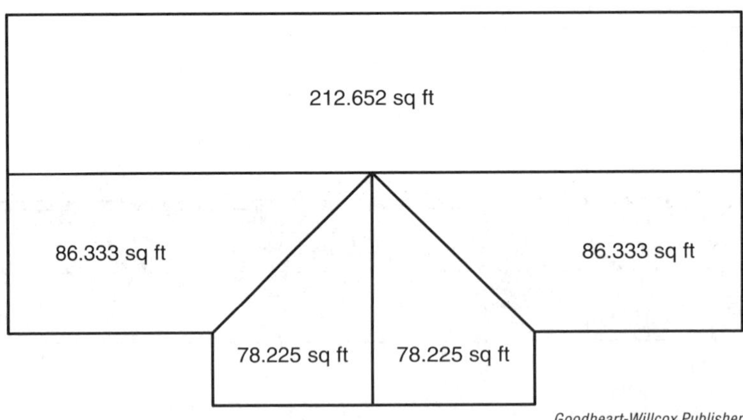

Goodheart-Willcox Publisher

Name ______________________ **Date** ____________ **Class** ____________

19. A general contractor received a payment of $15,000.00 following the completion of a project. After deducting the following expenses, what was the profit from the project?

    Materials: $5,365.98

    Equipment rental: $274.65

    Subcontractor: $3,287.75

    ____

20. Three windows were quoted to cost $787.99 by a local vendor. The cost of the first window was $287.56 and the second was $245.63. What was the cost of the third window?

    ____

21. What is the total length of the following board?

    ____

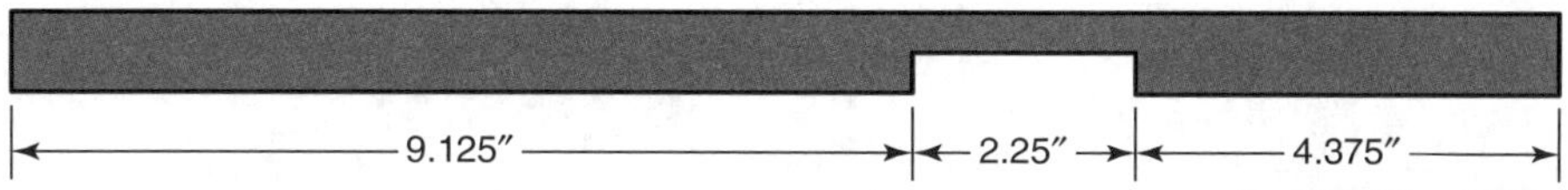

9.125″ | 2.25″ | 4.375″

Goodheart-Willcox Publisher

22. Five carpenters worked on a jobsite for a total of 100 hours. How many hours did Worker #5 work?

    Worker #1: 16.45 hours

    Worker #2: 27.15 hours

    Worker #3: 8.5 hours

    Worker #4: 32.333 hours

    Worker #5: ? hours

    ____

23. A contractor paid Juan $128.56, Scott $187.44, and Lindsey $155.75 for work they performed on a project. The contractor estimated $500.00 for the cost of labor. How much did the actual labor cost differ from his estimate?

    ____

24. If 24 squares of siding were delivered to a jobsite, and 14.57 squares were installed on the house and 7.68 squares on the garage, how many squares remain?

    ____

Name ______________________ **Date** ____________ **Class** ____________

What is the difference in elevations between each stake and the benchmark? Use the following image to answer questions 25–28.

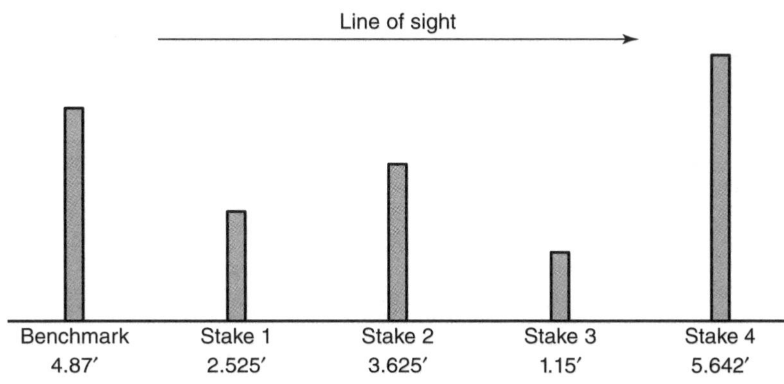

Goodheart-Willcox Publisher

25. Stake 1: __________

26. Stake 2: __________

27. Stake 3: __________

28. Stake 4: __________

29. A contractor received a $1,425.00 payment for a job. The materials costs were $523.87 and overhead for the job was $228.63. How much did the contractor keep as profit?

____

30. In the following image of a door threshold, calculate Distance A.

____

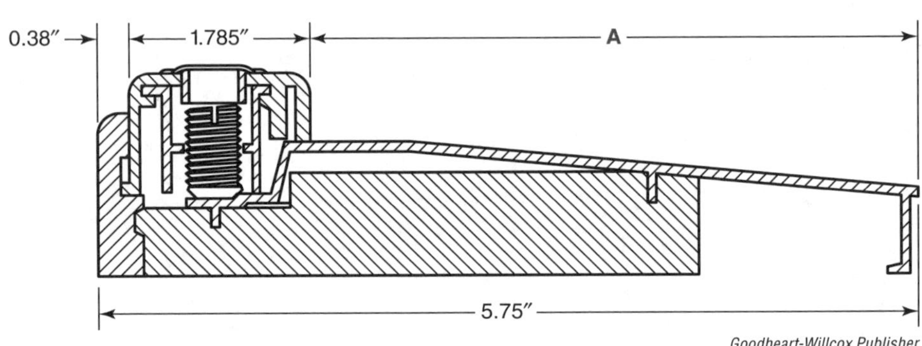

Goodheart-Willcox Publisher

UNIT 13

Multiplying Decimals

Objectives

After studying this unit, you will be able to:

- Demonstrate the method used to multiply decimals.
- Determine where to place a decimal in the product.

The process of multiplying numbers with decimals follows the same concepts that were covered in Unit 4, *Multiplying Whole Numbers*. The problems are set up and calculated the same way with the key difference being the placement of the decimal in the final answer.

Method Used to Multiply Decimals

Decimal numbers can be multiplied horizontally or vertically. Vertical multiplication aligns the numbers to the right allowing for simple multiplication and placement of the decimal in the final answer.

Horizontal multiplication:

$$42.7563 \times 12.6 = 538.72938$$

Vertical multiplication:

$$\begin{array}{r} 42.7563 \\ \times \quad 12.6 \\ \hline 538.72938 \end{array}$$

Horizontal multiplication will generally require a calculator to solve the problem and place the decimal. When multiplying decimal numbers, it is best to arrange the numbers vertically in a column aligned to the right side. There is no need to keep the decimals in-line for multiplication setup.

Incorrect	Correct
$\begin{array}{l} 16.854 \\ \times 6.3 \\ \hline \qquad ? \end{array}$	$\begin{array}{r} 16.854 \\ \times \quad 6.3 \\ \hline ? \end{array}$

Once the problem is set up correctly, the numbers are multiplied with no regard for the decimals.

Example 13-1

```
   42.7563
×     12.6
   2565378
   8551260
+ 42756300
         ?
```

Placing the Decimal in the Product

After the multiplication process is complete, the decimal point is placed in the product. To determine this placement, count the number of digits to the right of the decimal point in each of the numbers being multiplied. The decimal is placed in the product the same number of digits from the right.

Example 13-2

```
   42.7563  Four digits to the right of the decimal point
×     12.6  One digit to the right of the decimal point
   2565378
   8551260
+ 42756300
 538.72938
```

In total, there are five digits to the right of the decimal point in the numbers being multiplied. In the final answer, place the decimal point five digits from the right.

When multiplying decimals, be sure to multiply through all the digits, including zeros. Even though the products may be zero, these digits are needed for correct placement of the decimal.

Example 13-3

```
  0.0031  Four digits to the right of the decimal point
×    0.4  One digit to the right of the decimal point
 0.00124
```

There are five total digits to the right of the two numbers that were multiplied. After multiplying the zeroes, there are enough digits in the product to correctly place the decimal point.

Some multiplication problems may produce zeros at the end of the answer after the decimal point. If these zeros are the last digits to the right after the decimal point, they may be removed because they have no value.

Example 13-4

$$\begin{array}{r} 2200 \\ \times\ .45 \\ \hline 11000 \\ +\ 88000 \\ \hline 990.00 \end{array} = 990$$

Two digits to the right of the decimal point

Math Tip

When you are multiplying decimal numbers, always calculate a rough estimate to check that the decimal point is in the proper location. For your rough estimate, select numbers that are similar to the numbers being multiplied but that can also be multiplied together easily.

For example, if you are multiplying 183.47 × 21.53, you could estimate the answer by multiplying 200 × 20 to calculate an estimate of 4,000. The actual answer of 3,950.1091 is roughly approximate to the estimate. If you placed the decimal point in the wrong place and calculated 395.01091 or 39,501.091, the inconsistency of the product and your estimate would indicate an error in one of the calculations. Then you could go back and check your work or redo the problem.

Carpentry Notes

Checking the Diagonal of a Square

A concept that will be covered in great depth in Unit 21, *Right Angles* is determining the right angle or hypotenuse of a square. During that process, you must be able to multiply decimals and add them together. In the following layout, stakes one and two are spaced 12.5′ apart. Stake three is placed 10.5′ from stake one, but must also be the correct distance from stake two to ensure a right angle is created.

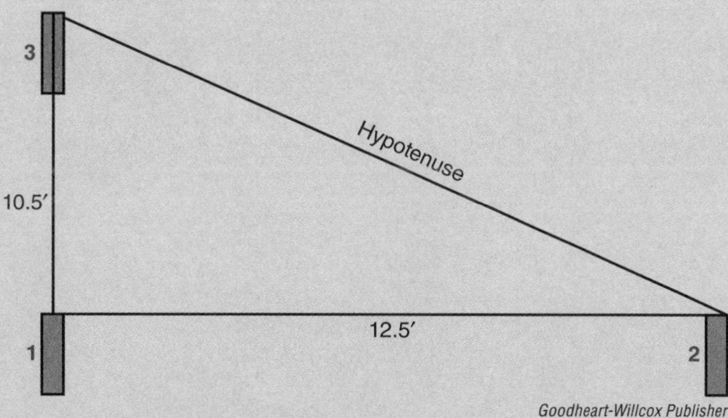

Goodheart-Willcox Publisher

To determine the hypotenuse length, the following formula is used:

$$a^2 + b^2 = c^2$$

$$12.5^2 + 10.5^2 = c^2 \text{ (multiply decimals)}$$

$$156.25 + 110.25 = c^2 \text{ (add decimals)}$$

$$266.5 = c^2$$

$$16.32482771731451' = c$$

Work Space/Notes

Unit 13 Review

Name ______________________ Date __________ Class __________

Solve the following equations. Make sure the decimal is correctly placed in your answer and show all of your work.

1. $2.5 \times 12.8 =$

2. $14.66 \times 32.07 =$

3. $22.125 \times 16 =$

4. $2.4 \times 0.875 =$

5. $31.005 \times 14 =$

6. $0.5215 \times 0.0075 =$

7. $\begin{array}{r} 0.72 \\ \times\, 0.39 \\ \hline \end{array}$

8. $\begin{array}{r} 0.618 \\ \times\ 7.48 \\ \hline \end{array}$

9. $\begin{array}{r} 12.6317 \\ \times\ \ 0.009 \\ \hline \end{array}$

10. $\begin{array}{r} 12.621 \\ \times\ 0.705 \\ \hline \end{array}$

11. $\begin{array}{r} 1.003 \\ \times\ 3.95 \\ \hline \end{array}$

12. $\begin{array}{r} 0.00685 \\ \times\ \ \ 0.24 \\ \hline \end{array}$

13. A wood project requires 14.75 bd ft of pine to construct. How many board feet are needed to build 12 of these projects?

____

14. A contractor hired a carpenter for $18.25 an hour to help complete a project. The carpenter worked a total of 32.5 hours. How much did the carpenter earn?

____

Name ______________________ Date ____________ Class ____________

15. Using the following diagram of a staircase, calculate the total rise of the stairs.

____

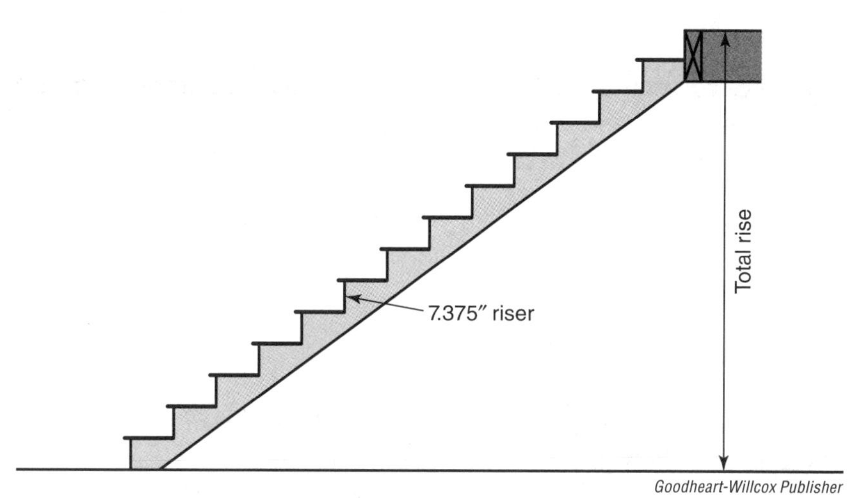

16. The flooring material needed for a room is estimated at 178.125 sq ft. The material sells for $3.79 per square foot. What is the total cost of flooring material for this room?

____

17. It takes an average of 2.25 hours for a carpenter to install and trim an interior door. How many hours will it take to install and trim 12 doors?

____

18. Based on the deck board sizes and spacing, what is the total width of the following exterior deck?

____

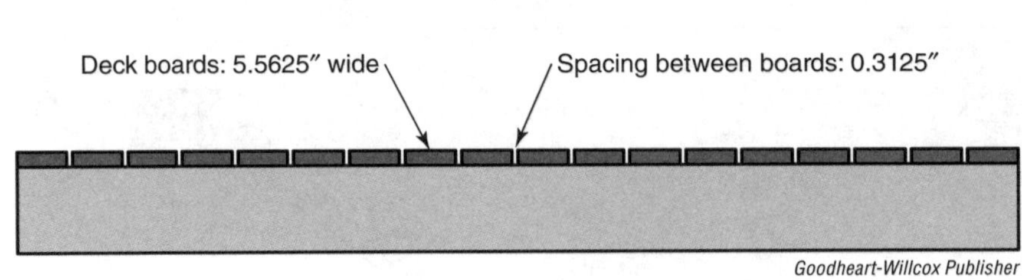

19. Wire closet shelving needs to be cut and installed in two linen closets. One closet requires 5 shelves cut to 18.5″, and the other closet requires 5 shelves cut to 22.75″. How much shelving material is needed to complete both closets?

____

20. Wood stock will be cut into lengths of 3.225′ for tread material for a set of utility steps. How much material is needed for 11 treads?

____

Name ______________________ Date ____________ Class ____________

21. What is the distance between post A and post B on the deck shown below?

____

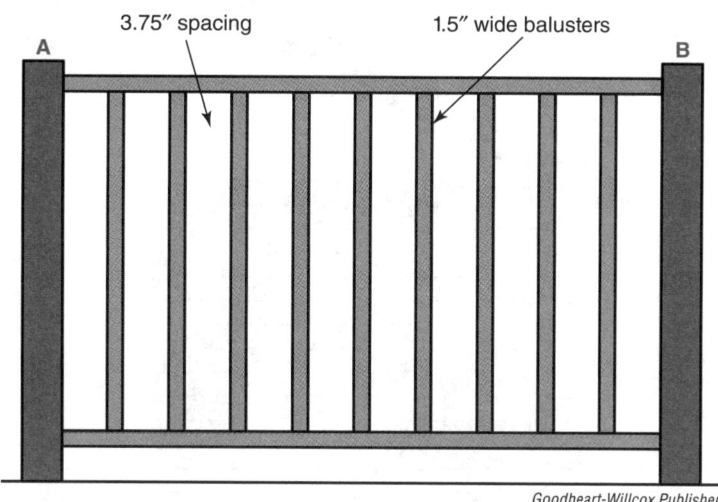

Goodheart-Willcox Publisher

22. A worker claiming mileage for his commute drives 26.8 miles round-trip every day. How many miles has he traveled in 15 workdays, and how much will he earn at $0.55 per mile?

____

23. A roofing contractor charges a base price of $325.50 per square installed. What would the estimated cost be for a roof requiring 18.25 squares of shingles?

____

24. What is the height of the log wall shown below?

____

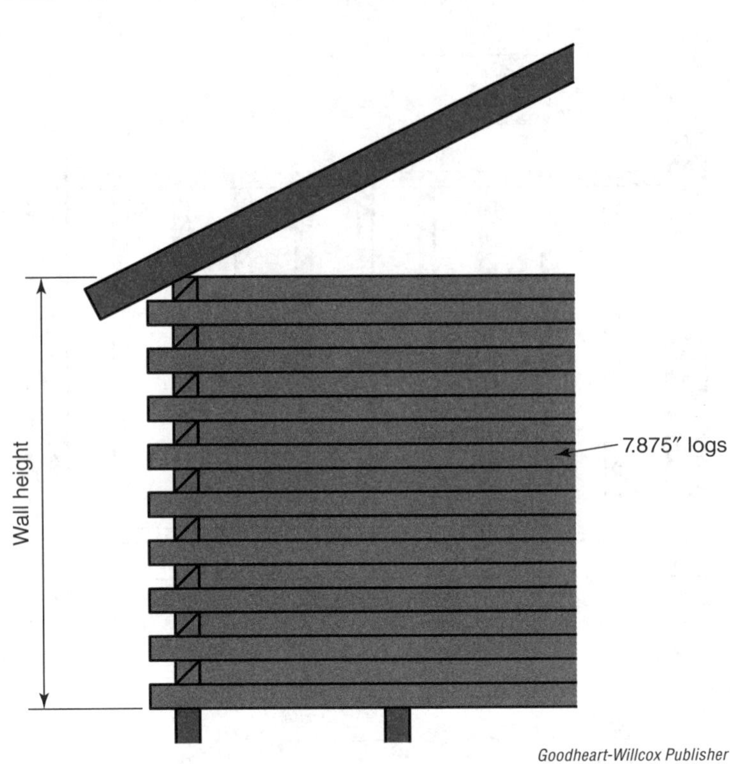

25. One bundle of three-tab shingles weighs 66.85 pounds. What will 16 squares of shingles weigh at 3 bundles per square?

____

UNIT 14

Dividing Decimals

Objectives

After studying this unit, you will be able to:

- Demonstrate the method used to divide decimals.
- Identify decimal placement within a quotient.
- Demonstrate how to convert fractions to decimals.

The process of dividing numbers with decimals follows the same concepts that were covered in Unit 5, *Dividing Whole Numbers*. While the problems are set up and calculated the same, adaptations must be made regarding placement of the decimal point in the equation.

Method Used to Divide Decimals

When dividing numbers with decimals, the numbers are set up within a division bracket as if they were whole numbers. The divisor must always be a whole number, but the dividend can be a decimal number when dividing.

$$24\overline{)165.302}$$

When both the dividend and the divisor are decimal numbers, division cannot begin until the decimal is eliminated from the divisor.

$$2.64\overline{)68.4972}$$

To remove the decimal from the divisor, move the decimal point to the right of the last digit within the divisor to create a whole number.

To keep the value of the equation the same, the decimal point in the dividend must then also be moved the same number of places to the right. After both decimal points have been moved, the equation can be completed.

$$264.\overline{)6849.72}$$

When there is no decimal point in the dividend, or not enough digits to move the decimal point, zeros can be added for proper placement of the decimal point.

Example 14-1

$$0.026\overline{)23.7} \quad = \quad 26\overline{)23700}$$

Math Tip

Always move the decimal point in the dividend the same number of places that it was moved in the divisor.

Decimal Placement in the Quotient

Once the decimal has been eliminated from the divisor, the division can be performed. If the dividend still contains a decimal, the quotient will have a decimal point placed in the same place value.

Example 14-2

$$0.4\overline{)1.04} \quad = \quad \begin{array}{r} 2.6 \\ 4\overline{)10.4} \\ \underline{-8} \\ 24 \\ \underline{-24} \\ 0 \end{array}$$

When there is a remainder leftover after dividing the divisor into the dividend, zeros can be added after the decimal point until the remainder is gone.

Example 14-3

$$1.6\overline{)10.7576} \quad = \quad \begin{array}{r} 6.7235 \\ 16\overline{)107.5760} \\ \underline{-96} \\ 115 \\ \underline{-112} \\ 37 \\ \underline{-32} \\ 56 \\ \underline{-48} \\ 80 \\ \underline{-80} \\ 0 \end{array}$$

Converting Fractions to Decimals

There are times when a common fraction must be converted to a decimal. In Unit 6, *Basic Principles of Fractions*, it was noted that the fraction bar has the same meaning as the division symbol. The denominator can be divided into the numerator, converting the fraction to a decimal.

Example 14-4

$$\frac{3}{8} = 3 \div 8 = \begin{array}{r} 0.375 \\ 8\overline{)3.000} \\ -0 \\ \hline 3\,0 \\ -2\,4 \\ \hline 60 \\ -56 \\ \hline 40 \\ -40 \\ \hline 0 \end{array}$$

There are some common fractions that, when converted to a decimal, will produce a decimal that is unending. This type of decimal will need to be rounded off.

Example 14-5

$$\frac{1}{3} = 1 \div 3 = .333333333333\ldots\ldots = .3$$

Carpentry Notes

Unit Rise Calculation

To determine the riser size in a set of stair stringers, the principles of converting a fraction to a decimal, dividing decimals, and converting a decimal to a fraction are all used.

A stairwell has a total rise of 106 7/8″. To determine the size of each riser in the set of steps, convert the total rise to a decimal.

$$\frac{7}{8} = 7 \div 8 = 0.875$$

$$0.875 + 106 = 106.875''$$

Divide the total stairwell height by 7″, the ideal riser height. Round the decimal to the nearest whole number to determine the number of risers.

$$106.875'' \div 7'' = 15.26785714285714 = 15 \text{ risers}$$

Divide the total height by the number of risers to find the size of each riser.

$$106.875'' \div 15 = 7.125''$$

Convert the riser size back to a fraction.

$$7.125 = 7 + \frac{125}{1000} = 7\frac{1}{8}''$$

Work Space/Notes

Unit 14 Review

Name ______________________ **Date** ____________ **Class** ____________

Divide the following decimal numbers, rounding all answers to three decimal places when needed. Show all of your work.

1. $0.4\overline{)69}$

2. 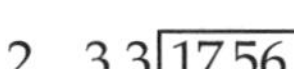 $3.3\overline{)17.56}$

3. $21\overline{)68.57}$

4. $0.02\overline{)20.001}$

5. $5.5\overline{)47.3}$

6. $0.45\overline{)117}$

7. $18\overline{)0.96}$

8. $6.6\overline{)87.479}$

9. $0.007\overline{)0.07}$

10. $24.76 \div 12.9 =$

Name ______________________ **Date** ____________ **Class** ____________

11. 95.302 cu yd ÷ 14 =

12. 0.0567 ÷ 11 =

13. 0.6481 ÷ 5.05 =

14. 9.09 ÷ 0.115 =

15. 85.75″ ÷ 13 =

Express the following common fractions as decimals.

16. $\frac{1}{8} =$

17. $\frac{9}{20} =$

18. $\frac{3}{8} =$

19. $\frac{4}{5} =$

20. $\frac{12}{48} =$

21. $\frac{6}{15} =$

Name ______________________ Date ____________ Class ____________

22. The total run of a set of stairs measures 143.5″. How many 10.25″ treads are in this set of stairs?

____

23. A furniture maker purchased 180 bd ft of unsurfaced red oak boards at a cost of $401.40. What was the cost of the oak per board foot?

____

24. How many 14.25″ boards can be cut from a plank measuring 131.75″ long?

____

25. A carpenter charged $22.50 per hour for a project that was bid at $681.25. If the material for the project is $400, how many hours of labor was included in the estimate?

____

26. A trim project requires 56 lineal feet of base molding. What is the price per lineal foot if the total cost was $97.44 for the material?

____

Use the roof shown to answer questions 27–28.

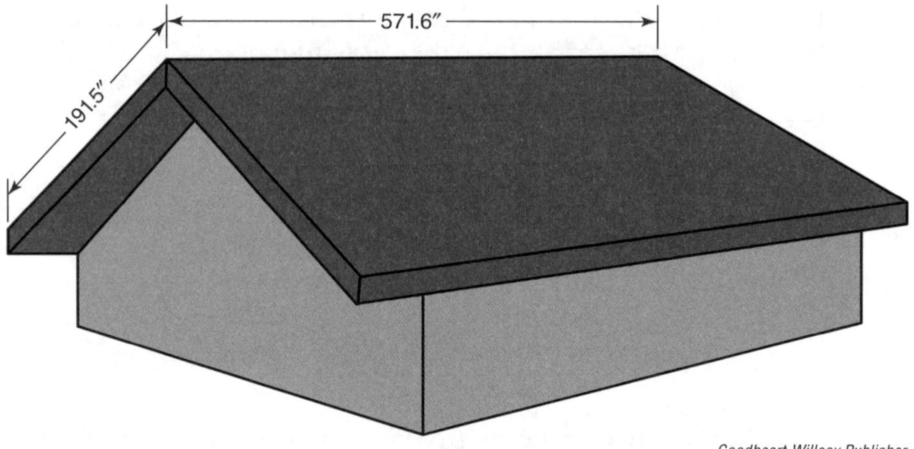

Goodheart-Willcox Publisher

27. Three-tab shingles measure 40″ long. How many shingles are needed to install one row of shingles across the gable roof?

____

28. The exposure on each row of three-tab shingles is 5.6″. How many rows of shingles are needed to complete one side of the roof?

____

Name ________________________ Date ____________ Class ____________

29. The total rise on a set of stairs measures 109.5″. How many 7.3″ risers are in this set of stairs?

____

The following storage rack is designed to hold full pieces of sheet goods of various thicknesses. Using the information in the table, determine how many full sheets can fit on each designated rack.

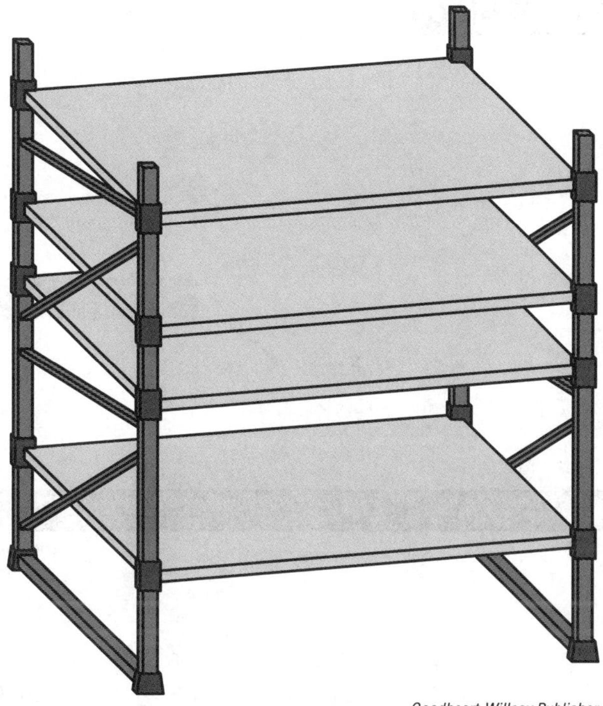

Goodheart-Willcox Publisher

	Thickness of One Sheet	Storage Space Allowance	Number of Sheets
30.	0.25″	14.75″	
31.	0.75″	42″	
32.	0.425″	21.25″	
33.	0.375″	16.875″	

Goodheart-Willcox Publisher

34. How many rows of 2.25″ hardwood flooring will it take to cover an office floor measuring 116.565″ wide?

____

35. Carpenters will be installing two sets of stringers in the stairwell shown below. How many 7.25″ risers will be in each stringer?

____

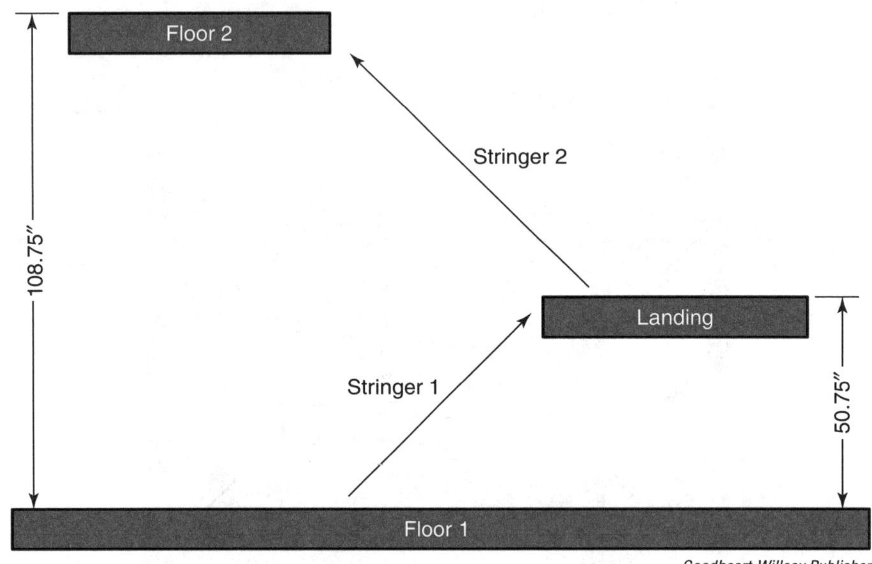

Goodheart-Willcox Publisher

*See Appendix A
Exterior Finish

36. An exterior wall measuring 99.75″ up to the ***soffit**** will be sided using ***horizontal wood siding**** with an exposure of 5.25″ per row. How many rows will it take to side the wall?

____

Name ______________________ Date ____________ Class ____________

37. The total rise on a set of steps measures 107 1/8″. What is this measurement in decimal form?

____

38. Two boards need to be surface planed to 1 inch thick. One measures 1 3/16″ and the other one measures 1.15625″. Which one needs to be planed more?

____

Work Space/Notes

Section 3 Exam

Name ______________________________ Date ____________ Class ____________

1. Round 0.47239 to the tenths place value.

   ____

2. Round 0.2098126 to the hundredths place value.

   ____

3. Round 0.0054891 to the thousandths place value.

   ____

4. Round 0.9007681 to the ten thousandths place value.

   ____

5. 284.63 + 289.04 =

6. 428.617 + 149.203 + 47.917 =

7. 271.13 – 159.04 =

8. 42.33 – 25.94 – 14.82 =

9. $5.1 \times 0.884 =$

10. $84.651 \times 12.1 =$

11. $$\begin{array}{r} 14.8516 \\ \times \quad 0.132 \\ \hline \end{array}$$

12. $$\begin{array}{r} 0.02415 \\ \times \quad 0.31 \\ \hline \end{array}$$

13. $97.35 \div 11.8 =$

14. $66.36 \div 6.32 =$

Name ______________________ **Date** __________ **Class** __________

15. Convert 3/16 to a decimal.

____

16. Convert 7/8 to a decimal.

____

17. Convert 11/16 to a decimal.

____

18. A window opening measures 32.375″ wide. What is this measurement in fraction form?

____

19. A room measures 127.125″ × 98.625″. What are the dimensions in fractional form?

____

20. A carpenter earned $2,364.25 on one job and $3,320.60 on another. After paying his helpers $1957.55 for their time, how much did he profit?

____

21. A contractor budgeted $1,200.00 to buy three new tools. The first two tools were purchased for $294.98 and $562.25 from a vendor. How much money is left to purchase the third tool?

____

22. Interior colonial casing sells for $1.89 per lineal foot. A carpenter needs 127 lineal feet for one job and 147 lineal feet for another job. How much will the total cost for casing be for both jobs?

____

23. A carpenter will be reimbursed for mileage at the rate of $0.55 per mile for an out-of-town job. The carpenter traveled 128 miles a day and worked for four days. How much was the carpenter paid for mileage?

____

Name ______________________ Date __________ Class __________

24. The total rise on a set of porch steps measures 43.5″. How many 7.25″ risers are in this set of stairs?

____

25. At the completion of a job, a carpenter was paid $1,200.00. The material for the job cost $360.00. At an hourly rate of $35.00 an hour, how many hours did the carpenter work?

____

Work Space/Notes

SECTION 4

Measurement

Key Terms

architect's scale
area
board foot
cube
dressed size
graduations
lineal measurement
nominal size
perimeter
polygon
rectangle
rectangular solid
rough cut
scale
square
triangle
US Customary system
volume

UNIT 15

Linear Measurement

Objectives

After studying this unit, you will be able to:

- Perform linear measurement using the US Customary system.
- Demonstrate the ability to read an architect's scale.
- Convert lineal measurements between units.
- Convert inches to decimal equivalents in feet.
- Convert decimal equivalents in feet to fractions.

The ability to read a rule and make accurate measurements is a fundamental skill required of all workers within the construction trades. **Lineal measurement** is the most basic form of measuring; it is defined as measuring in a straight line between two points. Carpenters must have a full understanding of the US Customary system and also be able to read measurements on a set of prints using an architect's scale. The ability to proficiently convert lineal measurements is necessary to ensure accurate calculations on site.

US Customary System

The **US Customary system** is the system of measure primarily used in the United States. Practically every task performed in the construction trades will require you to accurately measure using a standard measuring device, or a **scale**. Because of this, you must understand how the system works and become very proficient in the use of it.

The fundamental units within the scale are the inch, the foot, and the yard. There are 12 inches in a foot, 3 feet in a yard, and 36 inches in a yard.

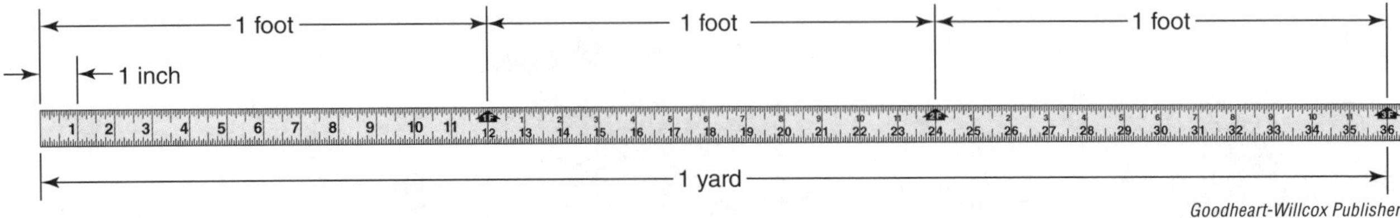

Goodheart-Willcox Publisher

Each inch within the scale is then broken up into fractional parts of an inch, or **graduations**, so the scale can be used to make precise measurements. Cutting the inch in two equal parts creates a 1/2″ line, and cutting the two halves equally creates 1/4″ lines. Cutting the quarters equally creates 1/8″ lines, and finally cutting the eighths in half creates 1/16″ lines. Some measuring instruments are graduated down further to 1/64″, but measuring in the construction trades generally is taken to 1/16″.

To help identify the value of the graduations on a rule, the lines vary in length. The whole inch marks are the longest with lines reducing in length down to 1/16″.

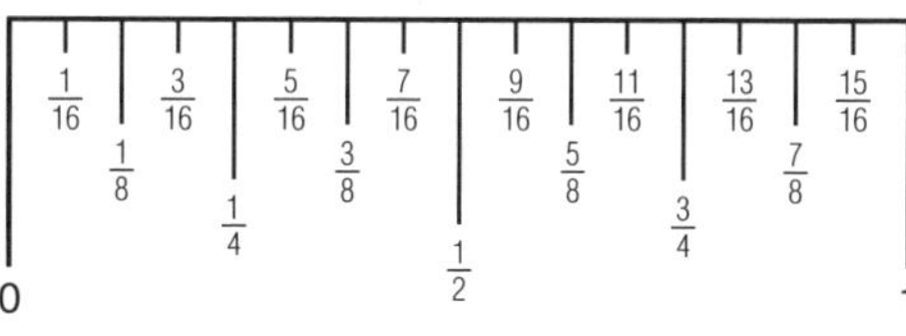

Goodheart-Willcox Publisher

Math Tip

All fractional inch measurements will be reduced to the lowest terms. The 1/16″ lines with even numbered numerators will always reduce to an eighth, quarter, or half. For example, 4/16″ = 1/4″.

When an object is measured with the scale, its length is determined by sequentially reading the graduations first in whole inches, determining the fractional value, and then combining the two.

Example 15-1

What is the value indicated below?

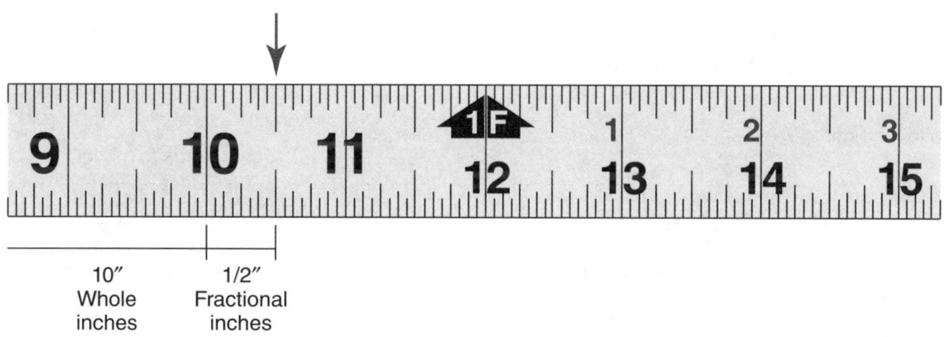

Goodheart-Willcox Publisher

Combine the whole and fractional inches to determine the value.

10″ + 1/2″ = 10 1/2″

When reading measurements within the scale, there are two ways to express the value of a measurement. On tape measures, variables of 12″ are also identified by their value in feet, and 12″ inches is identified as '1F' or 1 foot. The inch graduations that follow a whole foot are then identified by their value within the foot and are marked in red.

Goodheart-Willcox Publisher

As shown in the previous image, the value of the arrow is expressed as 14 3/8″. By using the foot and inch markings, it can also be expressed as 1′ 2 3/8″. This is a favorable method to use when working with larger numbers or when adding multiple measurements. For example, if material needs to be ordered to cut rafters measuring 174″ or 14′ 6″, it is easier to determine that 16′ material is needed based on the 14′ 6″ designation.

Reading an Architect's Scale

An **architect's scale** is a specialized device used on architectural drawings. Scales were used by architects to draw blueprints before the development of computer-aided design and drafting (CADD) programs. Architect's scales are still useful in determining lineal measurements on a set of prints in the field or for drawing projects to scale. Because a large construction project cannot easily be drawn full size on a blueprint, drawings are usually drawn to **scale**, a ratio between the actual size of a project and the size it is actually drawn. A drawing may have a scale of 1/4″ = 1′, meaning that 1/4″ in the drawing represents 1′ in the actual project. Since the scales vary in floor plans, plot plans, elevations, and details, the architect's scale has 11 different scales on its tri-shaped body.

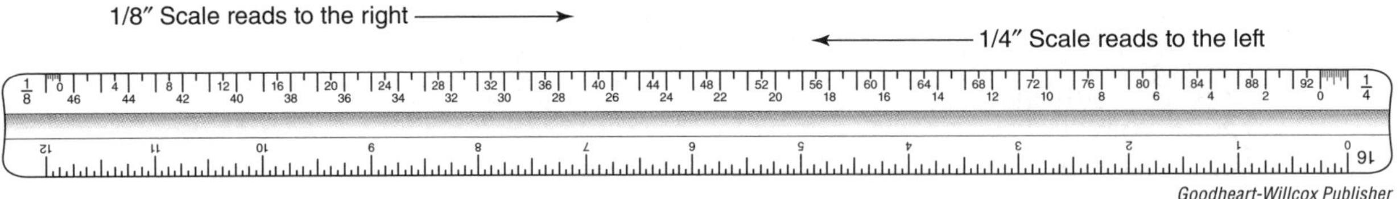

Goodheart-Willcox Publisher

One edge of the scale has a standard 12″ ruler graduated into 1/16″. The remaining edges of the scale rule have two rows of numbers running in opposite directions. One scale is read right to left and the other is read left to right. Each row of numbers has a zero at the end identified with the scale designation. Each individual scale is always read from zero toward the whole numbers, which represent whole feet. The graduated scale running the other direction from zero is always a whole foot graduated into fractions of a foot. Each scale has a different number of graduations.

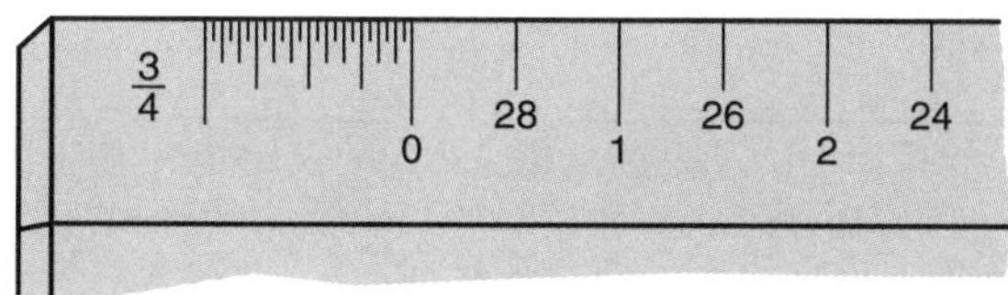

Goodheart-Willcox Publisher

The 3/4″ scale has a zero on the lower row of numbers indicating whole feet will be read left to right. The upper row of numbers must be ignored since they relate to the 3/8″ scale originating at the other end of the rule. The whole foot to the left of the zero is equivalent to 12″. Since there are 24 lines within that foot, each line is equivalent to a half inch.

To use the architect's scale to measure, first find the desired scale on the rule. The zero will identify what direction the scale is running and which of the two lines is to be read. The object to be measured is always measured by placing the zero at one end of the object and reading whole feet at the other. If the measurement falls between two whole feet, slide the scale back to the lower of the two feet and read the distance in feet and inches.

Example 15-2

What distance is identified on the scale?

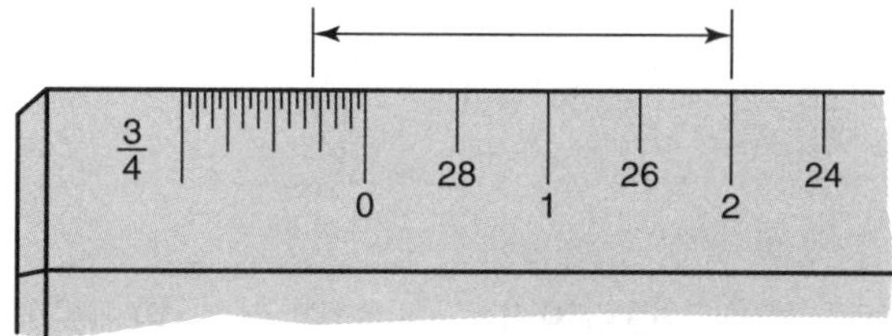

Goodheart-Willcox Publisher

Reading from the zero to the right, the whole feet reads 2′. On the 3/4″ scale, there are 24 lines to the left of the zero, meaning each line is equal to 1/2″. The scale measures 7 lines to the left of the zero.

$$7 \times 1/2'' = 3\ 1/2''$$
$$3\ 1/2'' + 2' = 2'\ 3\ 1/2''$$

Converting Lineal Measurements

Calculations within the construction industry will often require numbers to be converted from one unit of measure to another. To convert any number from feet to inches, simply multiply by 12, the number of inches in a foot.

Example 15-3

Convert 6′ to inches.

$$6' \times 12'' = 72''$$

When a dimension has both feet and inches, multiply the feet by 12 and add the additional inches.

Example 15-4

Convert 4′ 7″ to inches.

$$4' \times 12'' = 48''$$

$$48'' + 7'' = 55''$$

To convert any number from inches to feet, simply divide by 12.

Example 15-5

Convert 36″ to feet.

$$36'' \div 12'' = 3'$$

When converting from inches to feet, the number of inches being divided by 12 may not divide evenly leaving a remainder. If the answer can be expressed in feet and inches, the remainder simply becomes the inches.

Example 15-6

Convert 42″ to feet.

$$\begin{array}{r} 3\text{ r}6 = 3'\,6'' \\ 12\overline{)42} \\ -36 \\ \hline 6 \end{array}$$

The US Customary system uses fractions as parts of a whole but often mathematic calculations are performed using decimal fractions. Because of this, the ability to convert inches to decimal equivalents in feet and decimals to fractions by proportion is essential.

Converting Inches to Decimal Equivalents in Feet

Calculations may sometimes require you to convert inches to feet with the remainder being listed in decimal equivalents of a foot. Since there are 12 inches in a foot, any number of inches in the remainder divided by 12 will provide an answer in decimal equivalents of a foot.

Example 15-7

Convert 75″ to feet.

$$\begin{array}{r} 6.25 = 6.25' \\ 12\overline{)75.00} \\ -72 \\ \hline 30 \\ -24 \\ \hline 60 \\ -60 \\ \hline 0 \end{array}$$

Converting Decimal Equivalents in Feet to Fractions

Situations will arise when performing calculations when a lineal distance ends up in the form of a decimal foot and must be converted to a fraction of a foot so it can be measured with a tape measure.

Example 15-8

Convert 14.786241′ to fractional feet and inches.

Remove the whole feet from **14**.786241′.

$$0.786241'$$

To determine inches, multiply the decimal by 12″.

$$0.786241' \times 12'' = 9.434892''$$

Remove the whole inches from **9**.434892″.

$$0.434892''$$

To determine fractions of an inch, multiply the decimal by 16, the number of graduations in an inch on a standard tape measure.

$$0.434892 \times 16 = \mathbf{6.9}5$$

Because the tenth place value decimal is 5 or above, round up to the nearest whole number and place it in a fraction over 16 as the denominator. Add the whole feet and inches that were removed earlier to the fractional inch.

$$14.786241' = 14'\ 9\ 7/16''$$

Carpentry Notes

Converting Decimal Equivalents in Feet

Often in the construction trades, the diagonal measurements of a square or rectangle need to be calculated and then measured to check for square. This needs to be performed on walls, floors, decks, laying out corner stakes, etc. The process to determine the length of the diagonal measurement is covered in Unit 21, *Right Triangles*.

The wall shown has already had the diagonal measurement calculated, but the answer remains in the form of a decimal equivalent of a foot. Using the process to convert this number to a fractional foot will produce a measurement that can be found on a tape measure.

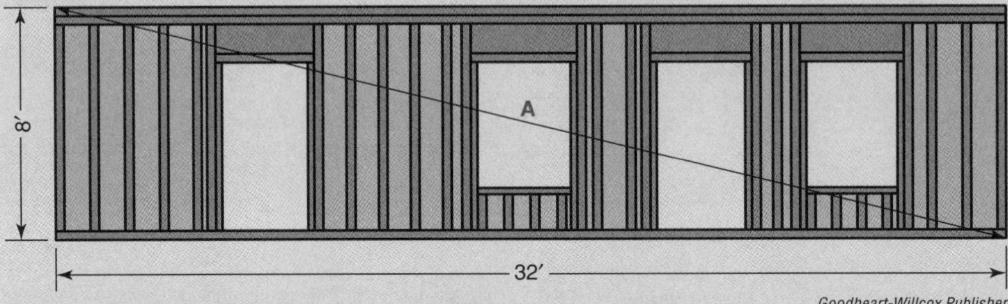

Goodheart-Willcox Publisher

Remove the whole feet from **32**.98484500494128′.

0.98484500494128′

To determine inches, multiply the decimal by 12″.

0.98484500494128 × 12″ = 11.81814005929541

Remove the whole inches from **11**.81814005929541.

0.81814005929541

Multiply the decimal by 16 to determine the fraction of an inch.

0.81814005929541 × 16 = 13.09024094872661

Round to the nearest whole number and place in a fraction over the denominator of 16.

13/16″

Add this fractional inch to the whole feet and inches that were removed earlier to determine the value of A.

32′ 11 13/16″

Unit 15 Review

Name ______________________ Date ____________ Class ____________

Identify the lineal measurements on the following rule.

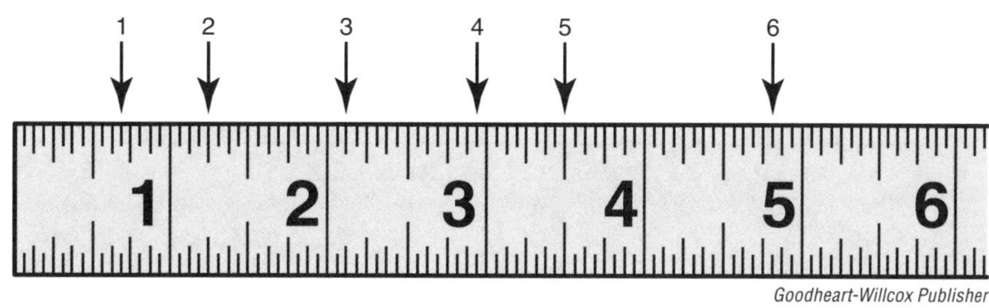

Goodheart-Willcox Publisher

1. ______________________
2. ______________________
3. ______________________
4. ______________________
5. ______________________
6. ______________________

Identify the lineal measurements on the following rule. Express your answers in feet and inches.

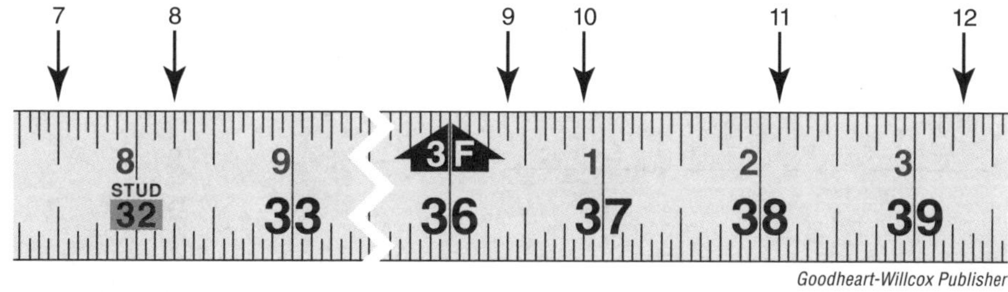

Goodheart-Willcox Publisher

7. ______________________
8. ______________________
9. ______________________
10. ______________________
11. ______________________
12. ______________________

Identify the scale measurements on the following architect's scales. Express your answers in feet and inches.

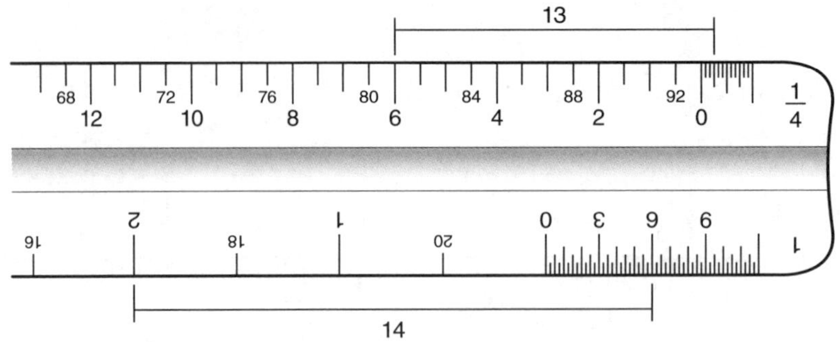

Goodheart-Willcox Publisher

13. ______________________

14. ______________________

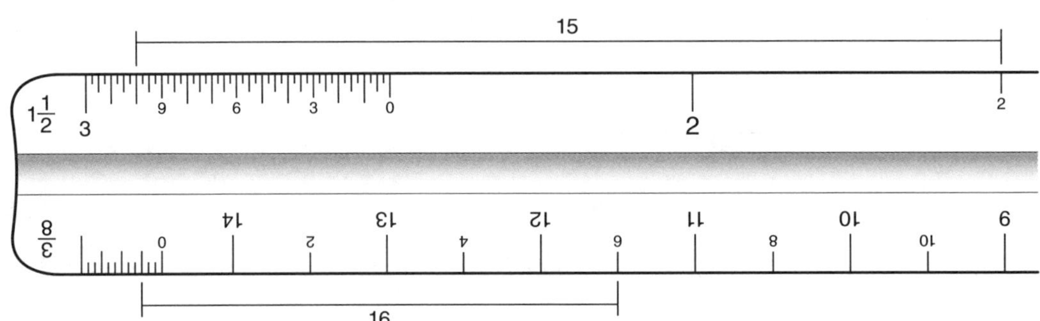

Goodheart-Willcox Publisher

15. ______________________

16. ______________________

Name ______________________________ **Date** ____________ **Class** ____________

Convert the following units of measure to inches.

17. 2′ 10″ =

18. 5′ 3″ =

19. 12′ 8″ =

20. 7′ 2 1/2″ =

21. 4′ 4 1/2″ =

Convert the following units of measure to feet and inches.

22. 73″ =

23. 41″ =

24. 109 3/4″ =

25. 57 1/16″ =

26. 29 7/8″ =

Name ______________________ **Date** __________ **Class** __________

Convert the following units of measure to decimal equivalents in feet. Round to the thousandths place value when needed.

27. 35″ =

28. 27″ =

29. 89″ =

30. 49″ =

31. 141″ =

Convert the following decimal equivalents in feet to fractional feet and inches.

32. 9.78531′ =

33. 11.28491′ =

34. 13.04937′ =

35. 10.8984′ =

36. 12.56471′ =

UNIT 16

Perimeter Measurement

Objectives

After studying this unit, you will be able to:

- Calculate the perimeter of a square.
- Calculate the perimeter of a rectangle.
- Calculate the perimeter of irregular shaped objects.

Purchasing materials for a project will often require you to calculate the **perimeter**, the distance around the outside of an object. The perimeter can be measured on any enclosed geometric figure, or **polygon**. A 14 sq ft room can have its perimeter measured for base molding. A rectangular foundation could be measured for sill plates, or an irregular shaped roof could be measured for drip edge. Perimeters are usually measured in lineal units of feet and inches.

Perimeter of a Square

A **square** is a four-sided polygon that has equal measurements on all four sides and four 90° corners. Using the side length (s), two formulas can be used to determine a square's perimeter (P).

$$P = s + s + s + s$$

$$P = 4 \times s$$

Example 16-1

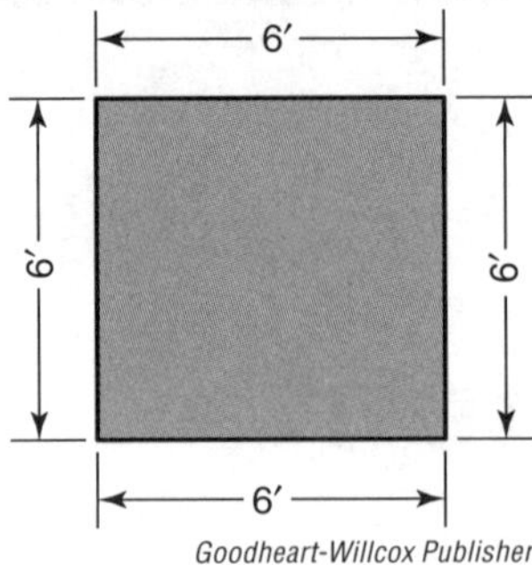

Goodheart-Willcox Publisher

When determining the perimeter of a square, either formula can be used.

$P = s + s + s + s$

$P = 6' + 6' + 6' + 6'$

$P = 24'$

$P = 4 \times s$

$P = 4 \times 6'$

$P = 24'$

The perimeter of the square is 24′.

Perimeter of a Rectangle

A **rectangle** is a polygon with four sides whose opposite sides are both parallel and equal in length. Like a square, its corners are 90° and two formulas can be used to determine a rectangle's perimeter.

$P = \text{length } (l) + \text{width } (w) + \text{length } (l) + \text{width } (w)$

$P = (2 \times l) + (2 \times w)$

Example 16-2

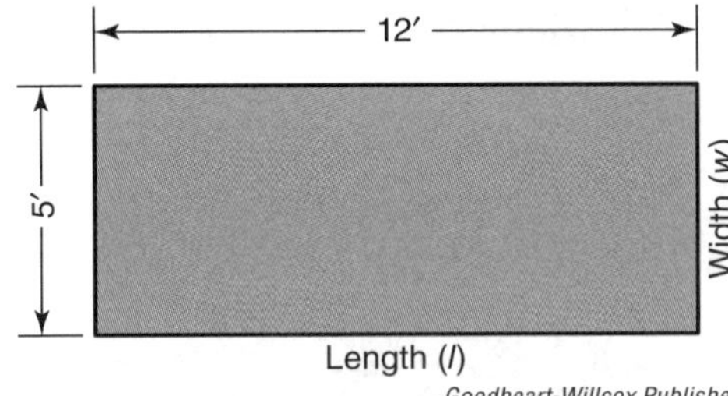

Goodheart-Willcox Publisher

When determining the perimeter of a rectangle, either formula can be used.

$P = l + w + l + w$

$P = 12' + 5' + 12' + 5'$

$P = 34'$

$P = (2 \times l) + (2 \times w)$

$P = (2 \times 12') + (2 \times 5')$

$P = 24' + 10'$

$P = 34'$

The perimeter of the rectangle is 34′.

Perimeter of Irregular Shaped Objects

Irregular polygons are common in construction. Not every room in a house, or roof on a building, will be in the shape of a square or rectangle. Because an object is irregular, a standard formula cannot be followed to determine the perimeter. All the sides of the irregular shaped object must be determined and then added together.

Example 16-3

To determine the perimeter of the room shown below, determine the length of all the sides and add them together. All corners of the room are square.

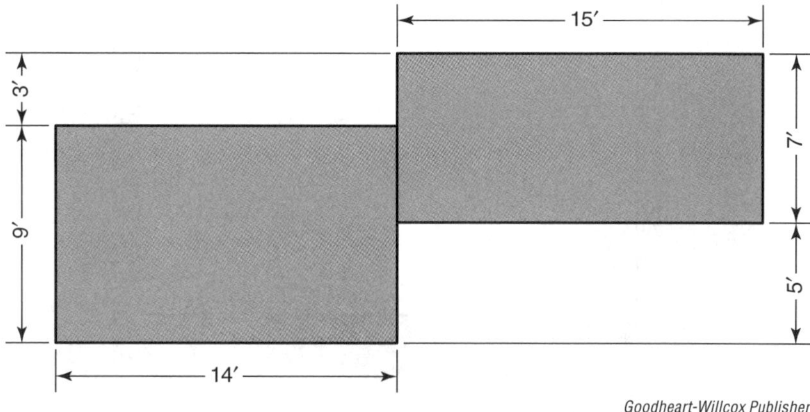

Goodheart-Willcox Publisher

The length of the walls running parallel to 14′ and 15′ walls are not identified, but since all the corners are square, they will also be 14′ and 15′ long. Be sure to account for every wall of the room.

$$P = 9' + 14' + 3' + 15' + 7' + 15' + 5' + 14'$$

$$P = 82'$$

The perimeter of the room is 82′.

Math Tip

When performing perimeter calculations, all measurements must be expressed in the same unit. Measurements expressed as feet and inches (8′ 3″) are often converted to a decimal form (8.25′) for calculating purposes.

Carpentry Notes

Sill Plate Estimation

*See Appendix A
Foundation and Floor Frame

Once the footer and foundation of a building have been constructed, the sill plate is installed on top of the foundation and secured with ***anchor bolts****. The sill plate allows a wooden structure to be connected to the concrete or block foundation.

To determine the number of plates needed, start by determining the perimeter of the following foundation.

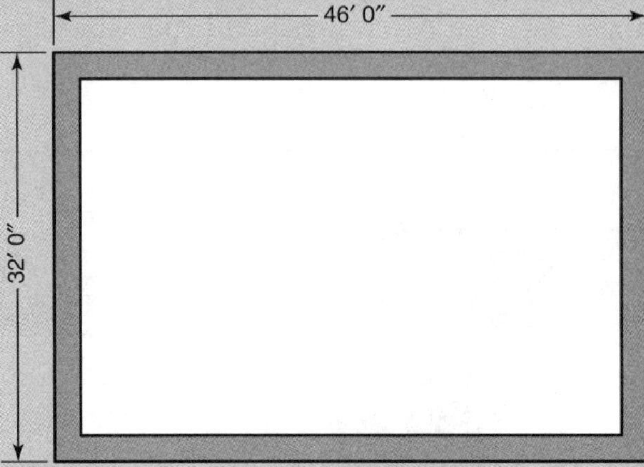

Goodheart-Willcox Publisher

$$P = (2 \times l) + (2 \times w)$$
$$P = (2 \times 46') + (2 \times 32')$$
$$P = 92' + 64'$$
$$P = 156'$$

After the perimeter has been found, divide by the length of sill plate stock (16′).

$$156' \div 16' = 9.75$$

Ten 16′ plates are needed.

Unit 16 Review

Name ______________________ Date ____________ Class ____________

Solve the following perimeter problems. Show all of your work.

1. What is the perimeter of a garage that is 24′ wide and 24′ long?

   ____

2. What is the perimeter of a house that is 32′ wide and 66′ long?

   ____

3. How many lineal feet of fencing is needed to fence in the perimeter of a yard that is 60′ wide and 155′ long?

   ____

4. How many 10′ pieces of drip edge are needed to trim the entire perimeter of the roof shown below?

____

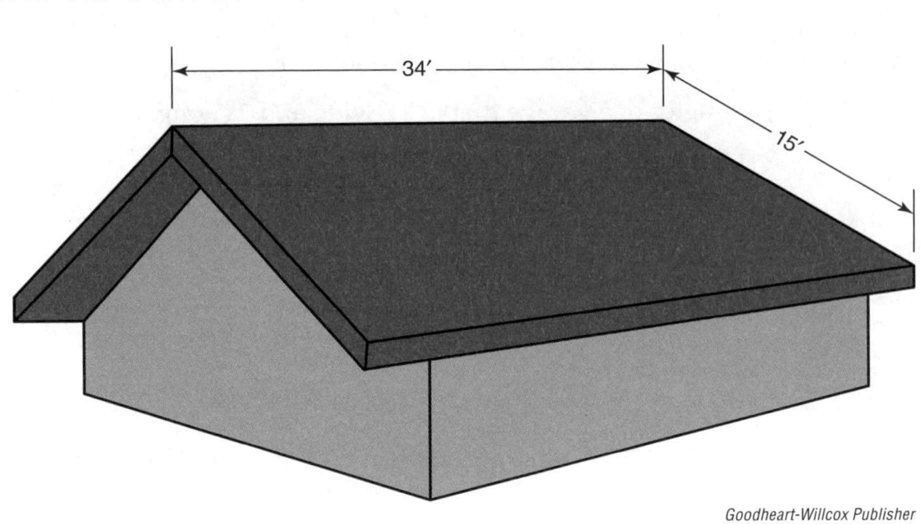

Goodheart-Willcox Publisher

5. What is the perimeter of a room that is 12′ 9″ wide and 14′ 3″ long?

____

6. What is the perimeter of a closet that measures 63 1/2″ long and 24 1/4″ wide? Express your answer in feet and inches.

____

Name ______________________ Date ____________ Class ____________

7. How many 12′ pieces of wall angle are needed to install around the perimeter of a room measuring 52′ 7″ × 48′ 10″? (Note: Wall angle is installed around the perimeter of a room at ceiling height and is part of a suspended ceiling grid system.)

____

Use the following foundation wall diagram to answer questions 8–9.

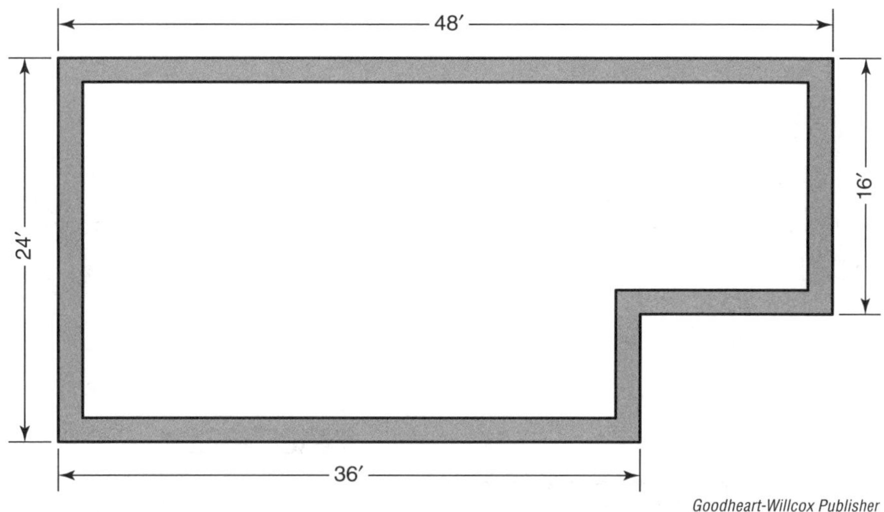

Goodheart-Willcox Publisher

8. How many lineal feet of sill seal are needed to seal the foundation wall?

____

9. How many 14′ sill plates are needed to plate the foundation?

____

10. A bedroom measures 14′ 6″ × 16′ 9″ and has a walk-in closet measuring 8′ 4″ × 8′ 8″. For ordering baseboard, calculate the combined perimeter of the bedroom and closet in feet and inches.

____

11. Fence posts will be evenly spaced every 8′ around the perimeter of a yard to enclose a play area measuring 24′ 0″ × 24′ 0″. How many posts will be needed?

____

*See Appendix A
Interior Finish

12. How many lineal feet of ***crown molding**** will be needed for the room shown in the following image? Round your answer to the nearest foot.

____

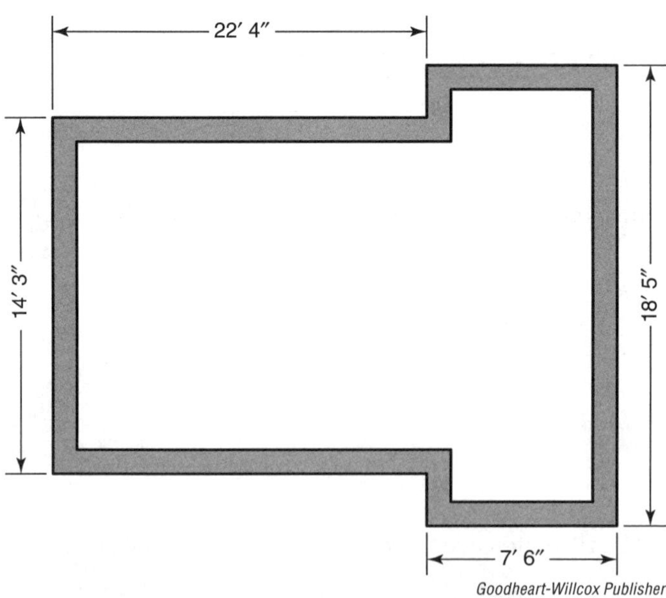

Goodheart-Willcox Publisher

Name ______________________ Date ____________ Class ____________

13. A 22′ 6″ × 20′ 9″ room needs to have carpet tack strips nailed around the perimeter of the room. How many tack strips are needed if each tack strip is 4′ long? (Note: Tack strips are installed on the floor around the perimeter of a room to secure the carpeting when installed.)

____

Use the following image to answer questions 14–15.

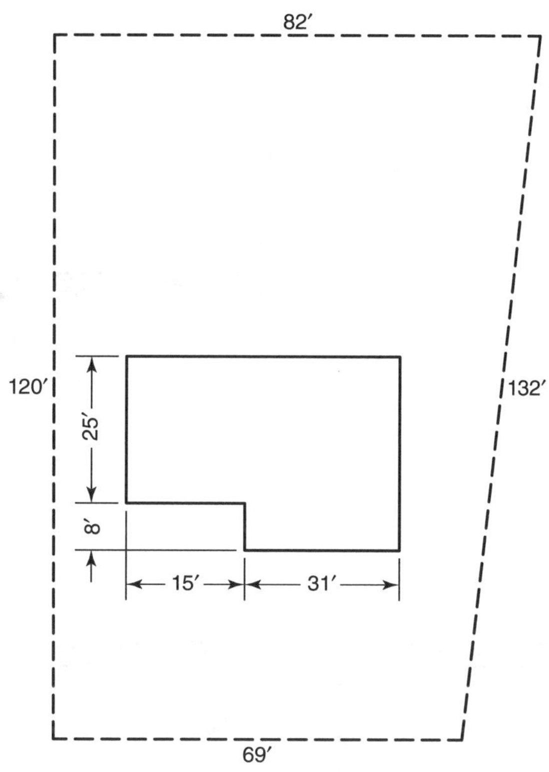

Goodheart-Willcox Publisher

14. What is the perimeter of the property?

____

15. What is the perimeter of the house shown on the plot plan?

____

16. An island countertop measures 78 3/4″ × 36 5/8″. What is the perimeter of the island in feet and inches?

____

Use the following outline to answer questions 17–18.

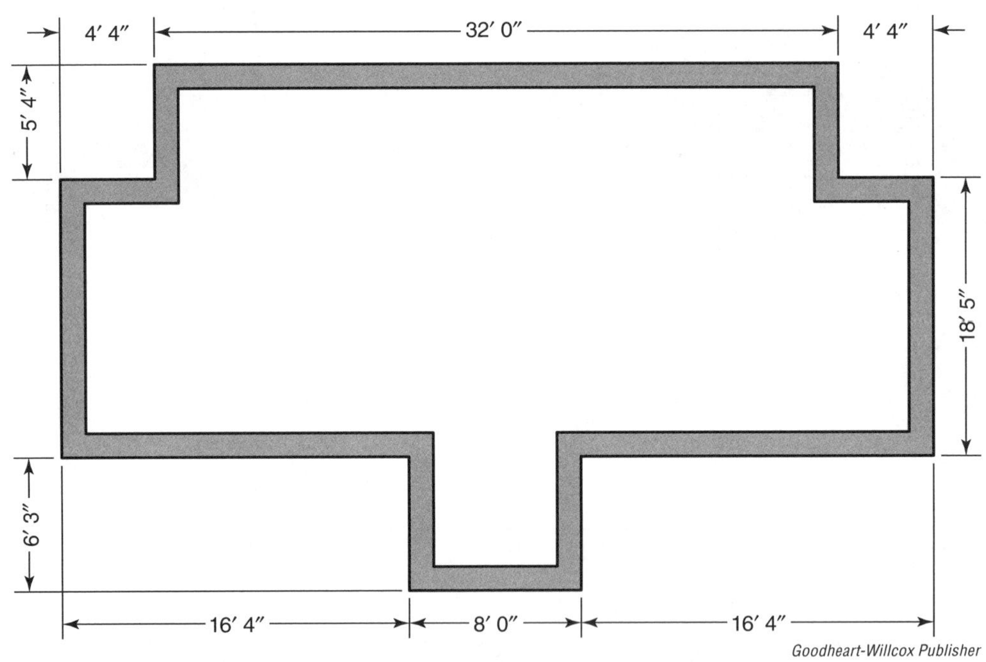

Goodheart-Willcox Publisher

17. What is the perimeter of the house in feet and inches?

____

Name ______________________ Date ____________ Class ____________

18. How many 14′ sill plates are needed for the exterior walls?

____

19. A ***hip roof**** measures 49′ 4″ × 27′ 4″. How many 12′ pieces of aluminum fascia are needed to cover the sub fascia along the perimeter of the roof?

____

*See Appendix A
Roof Types and Terminology

20. What is the perimeter of a room that is 15′ 6″ long and 114″ wide?

____

21. How many lineal foot of ***chair rail**** is needed for the perimeter of a dining room measuring 15′ 6″ × 18′ 3″?

____

*See Appendix A
Interior Finish

22. What is the perimeter of a kitchen that measures 14′ 9″ × 165″?

____

23. The perimeter of a square room is 58′ 0″. What is the length of the four walls?

____

Refer to the following building to answer questions 24–25.

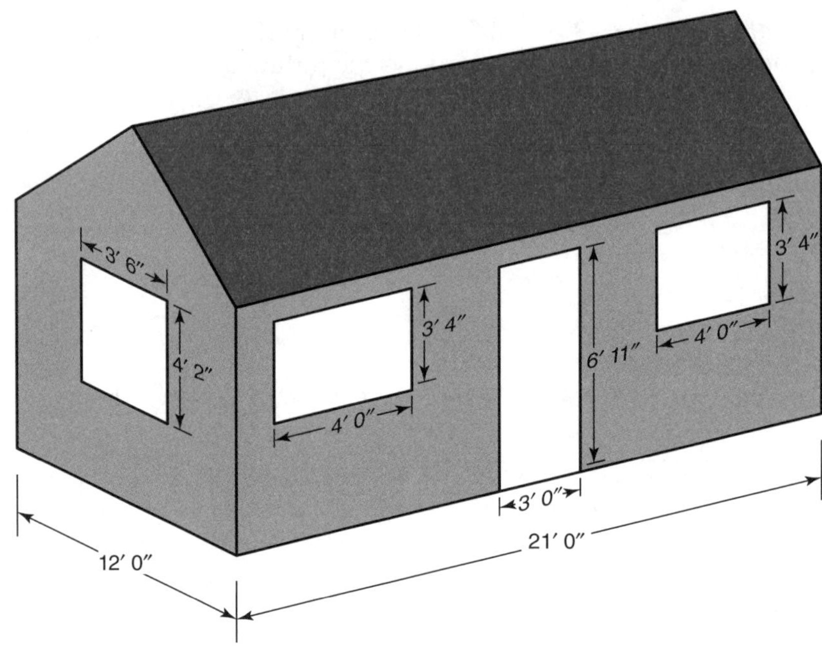

Goodheart-Willcox Publisher

*See Appendix A
Exterior Finish

24. How many lineal foot of ***starter strip**** is needed around the perimeter of the building for siding installation? (Note: Do not include the door opening.)

____

*See Appendix A
Exterior Finish

25. When applying siding, ***J-channel**** is installed around the top and sides of all windows and doors. How many lineal foot of J-channel is needed?

____

UNIT 17

Area Measurement

Objectives

After studying this unit, you will be able to:

- Calculate the area of a square.
- Calculate the area of a rectangle.
- Calculate the area of a triangle.
- Calculate the area of irregular shaped objects.
- Convert square units of measure.

In Unit 16, *Perimeter Measurement*, perimeter was defined as the distance around the outside of a polygon. **Area** is defined as the amount of surface within a polygon. Area is measured in square units, such as square inches, square feet, or square yards.

Throughout the building process, the surface area of squares and rectangles must be calculated to determine material quantities, such as sheathing, siding, or roofing material. With the complexity of many architectural designs, the area of irregular shapes must be measured for material quantities. This is done by breaking the shape into squares, rectangles, and triangles, measuring them individually, and then adding them together to find the total.

It is often necessary to divide an area into known shapes to determine surface area. To determine material quantities for the gable end of a house, the wall may be divided into a rectangle and a triangle. Unlike squares and rectangles that consist of four sides, a **triangle** is a polygon that is formed by only three sides.

Zern Liew/Shutterstock.com

Before calculating area, all measurements must be expressed in the same unit. After an area measurement has been determined, square units of measure can then be converted from one unit to another.

Area of a Square

To determine the area of a square, multiply both sides together and express the answer in square units.

Example 17-1

The area of an 8″ square is 64 sq in.

$$A = s \times s$$
$$A = 8'' \times 8''$$
$$A = 64 \text{ sq in}$$

When calculating the area of a square containing denominate numbers, their units combine to create a squared unit. Area is then expressed in square units using a variety of notations. Square inches are often expressed as "sq in" or "in^2."

Area of a Rectangle

To determine the area of a rectangle, multiply the length by the width and express the answer in square units.

Example 17-2

The area of a 16″ × 5″ rectangle is 80 sq in.

$$A = l \times w$$
$$A = 16'' \times 5''$$
$$A = 80 \text{ sq in}$$

Math Tip

When performing area calculations, all measurements must be expressed in the same unit. A rectangle measuring 15″ × 2′ would be calculated as either 15″ × 24″ or as 1.25′ × 2′.

Area of a Triangle

Before determining the area of a triangle, the parts of a triangle must be identified. The base and height of a triangle run perpendicular to one another. The base is the horizontal side the triangle sits on and the height is the length from the base to the highest point of the triangle.

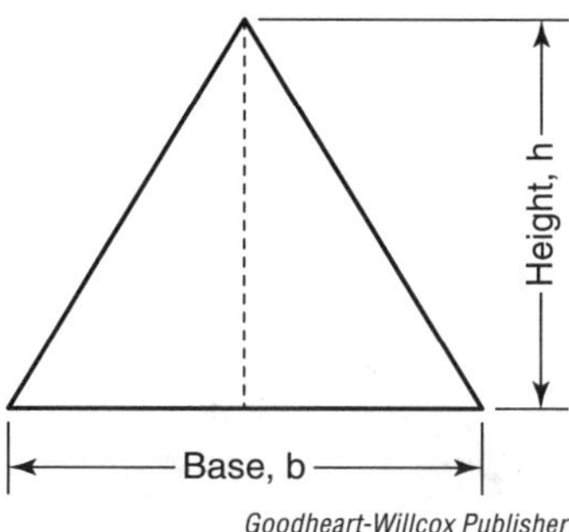

Goodheart-Willcox Publisher

To determine the area of a triangle, find 1/2 of the base (b) × the height (h) and express the answer in square units.

Example 17-3

Determine the area of a triangle with a 12″ base and a 10″ height is 60 sq in.

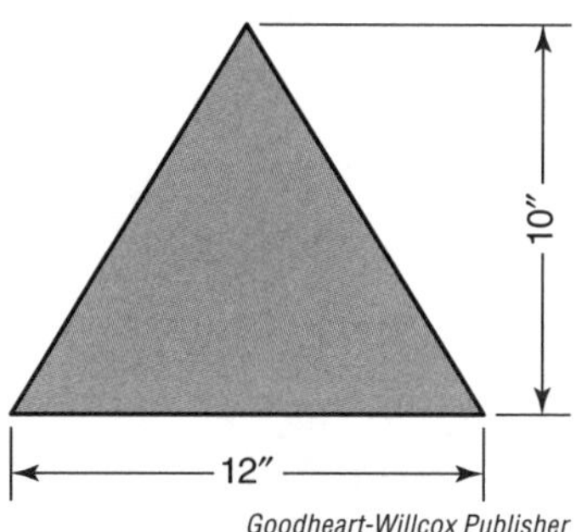

Goodheart-Willcox Publisher

$$A = 1/2\ (b \times h)$$
$$A = 1/2\ (12'' \times 10'')$$
$$A = 1/2\ (120 \text{ sq in})$$
$$A = 60 \text{ sq in}$$

Area of Irregular Shaped Objects

To determine the area of irregular shaped objects, the shape must first be broken down into squares, rectangles, and triangles as needed. The areas of the individual shapes are first calculated and then added together to determine the total area of the irregular shaped object.

Example 17-4

To calculate the area of the side of a house with a gable roof, break the wall into a rectangle and a triangle.

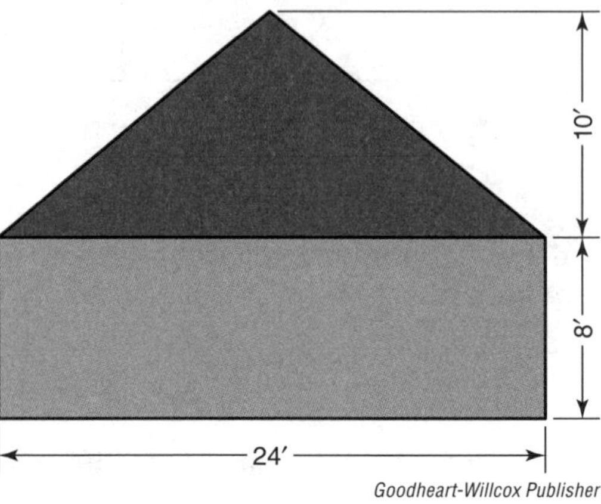

Goodheart-Willcox Publisher

Calculate the area of the rectangle.

$$A = l \times w$$
$$A = 24' \times 8'$$
$$A = 192 \text{ sq ft}$$

Calculate the area of the triangle.

$$A = 1/2\,(b \times h)$$
$$A = 1/2\,(24' \times 10')$$
$$A = 1/2\,(240 \text{ sq ft})$$
$$A = 120 \text{ sq ft}$$

Add the areas of the rectangle and triangle to determine the area of the entire wall.

$$192 \text{ sq ft} + 120 \text{ sq ft} = 312 \text{ sq ft}$$

Converting Square Units of Measure

In Unit 15, *Linear Measurement*, the practice of converting lineal measurements to different types of units was demonstrated. The same type of conversions need to be made when performing area calculations. For example, carpeting is sold by the square yard, so the area calculations for a room to be carpeted would need to be converted from square feet to square yards.

To make the conversions between square inches, feet, and yards, you must first know how many square inches are in a square foot and how many square feet are in a square yard.

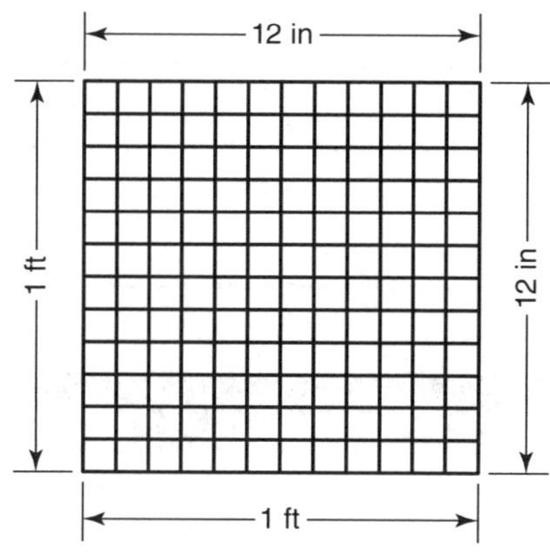

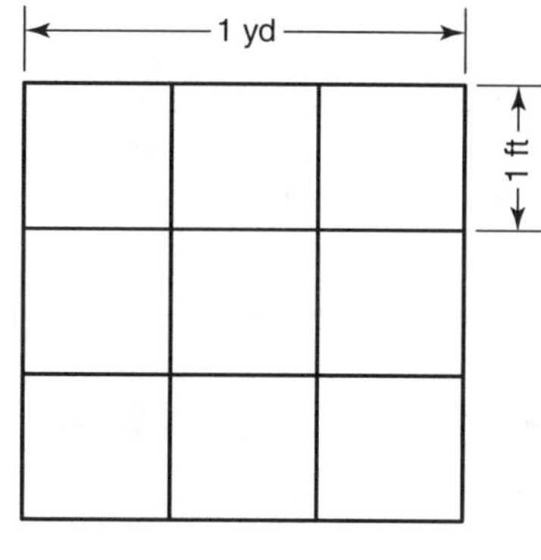

Goodheart-Willcox Publisher

A foot measures 12″, and a square foot measures 12″ by 12″. When applied to the area formula, a square foot contains 144 square inches.

$$1 \text{ sq ft} = 12'' \times 12'' = 144 \text{ sq in}$$

A yard can also be expressed as 3′. A square yard measuring 3′ by 3′, applied to the square feet formula, would then contain 9 square feet.

$$1 \text{ sq yd} = 3' \times 3' = 9 \text{ sq ft}$$

To make conversions then between square units, the following formulas are used:

$$\text{sq in} = \text{sq ft} \times 144$$
$$\text{sq ft} = \text{sq in} \div 144$$
$$\text{sq ft} = \text{sq yd} \times 9$$
$$\text{sq yd} = \text{sq ft} \div 9$$

Example 17-5

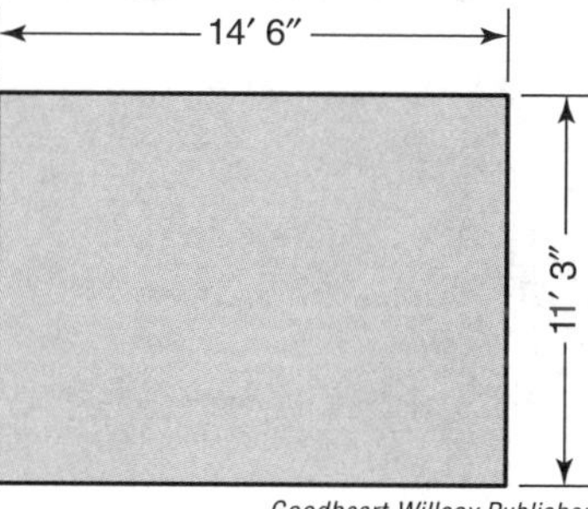

Goodheart-Willcox Publisher

Using the conversion formula, carpet is measured for a room in square feet and then converted to cubic feet.

Calculate the area of the room in square feet.

$$A = 14'\,6'' \times 11'\,3''$$
$$A = 14.5' \times 11.25'$$
$$A = 163.125 \text{ sq ft}$$

Convert from square feet to cubic yards.

$$163.125 \text{ sq ft} \div 9 = 18.125 \text{ sq yd of carpet needed}$$

Math Tip

To calculate the amount of sheet goods needed, the total area to be sheathed must be divided by the area of one sheet. A majority of sheet panel products measure $4' \times 8'$ so they contain 32 sq ft. An area of 480 sq ft, for example, will need 15 sheets.

$$480 \text{ sq ft} \div 32 = 15 \text{ sheets required}$$

Carpentry Notes

Subfloor Estimation

When building a house, the floors must be sheathed before building the exterior walls and partitions. To determine the amount of subflooring required, the area of the floor must be calculated, and then the square footage is divided by the area covered by one sheet of flooring, 32 sq ft.

How many sheets of 3/4″ T & G sheathing are needed for the following floor system?

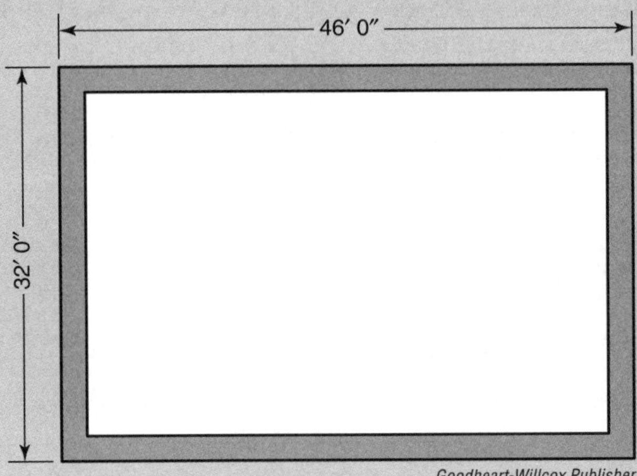

Goodheart-Willcox Publisher

$$A = l \times w$$
$$A = 32'\ 0'' \times 46'\ 0''$$
$$A = 1{,}472 \text{ sq ft}$$
$$1{,}472 \text{ sq ft} \div 32 = 46 \text{ sheets required}$$

Unit 17 Review

Name ______________________________ **Date** ____________ **Class** ____________

Determine the area of the following polygons. Show all of your work.

1. Square with 9″ sides.

   ____

2. Square with 8′ 3″ sides.

   ____

3. Rectangle with a length of 12″ and a width of 8″.

   ____

4. Rectangle with a length of 7′ 10″ and a width of 6′ 4″.

____

5. Triangle with a 14′ base and a 5′ height.

____

6. Triangle with an 18′ 6″ base and a 9′ 9″ height.

____

Name ____________________ Date __________ Class __________

Refer to the following house to answer questions 7–10.

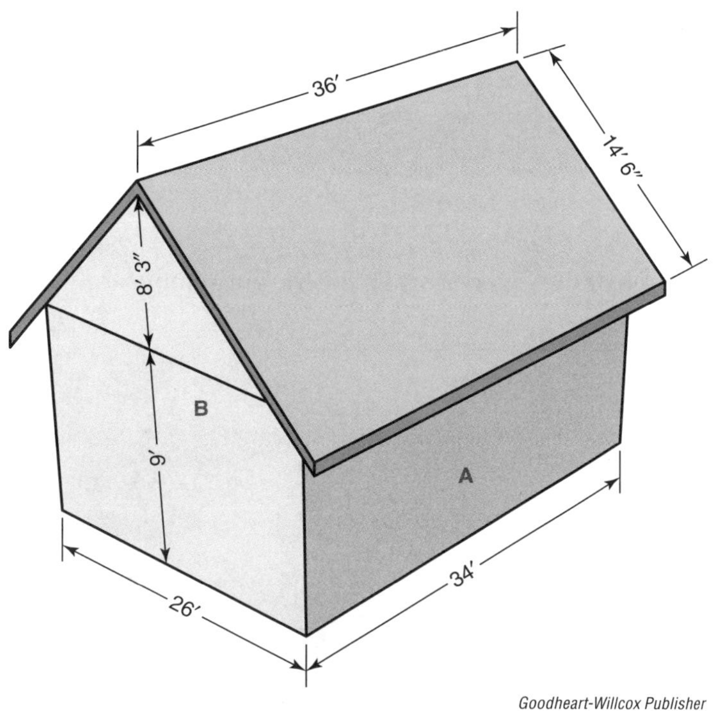

Goodheart-Willcox Publisher

7. What is the area of wall A?

____

8. What is the area of the entire gable roof?

____

9. What is the area of wall B?

____

10. What is the area of the exterior walls for the entire house?

____

The following floor plan will be used to answer questions 11–14.

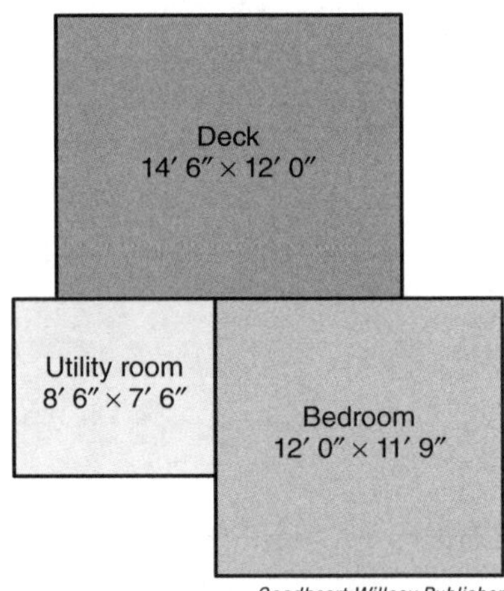

Goodheart-Willcox Publisher

11. What is the area of the bedroom?

____

Name ______________________________ Date ______________ Class ______________

12. What is the area of the utility room?

____

13. What is the area of the deck?

____

14. What is the total area of interior walls, if the walls are 9′ tall?

____

Convert the following area dimensions. Round your answers to two decimal places.

15. 365 sq in = _________ sq ft

16. 16 sq yds = _________ sq ft

17. 5.6 sq ft = _________ sq in

18. 487 sq ft = _________ sq yd

19. 15,552 sq in = _________ sq yd

20. 2.6 sq yd = _________ sq in

21. Calculate the flooring area in a 28′ × 52′ two-story house with a one-story 16′ × 14′ 6″ addition.

____

Name ______________________ Date ____________ Class ____________

22. How many 4′ × 8′ tongue-and-groove sheets of subfloor are needed to sheath a 30′ × 68′ floor system?

____

23. How many sheets of 4′ × 8′ roof sheathing are needed to sheath a 28′ long gable roof with a 13′ 4″ slope?

____

The following wall will be used to answer questions 24–26.

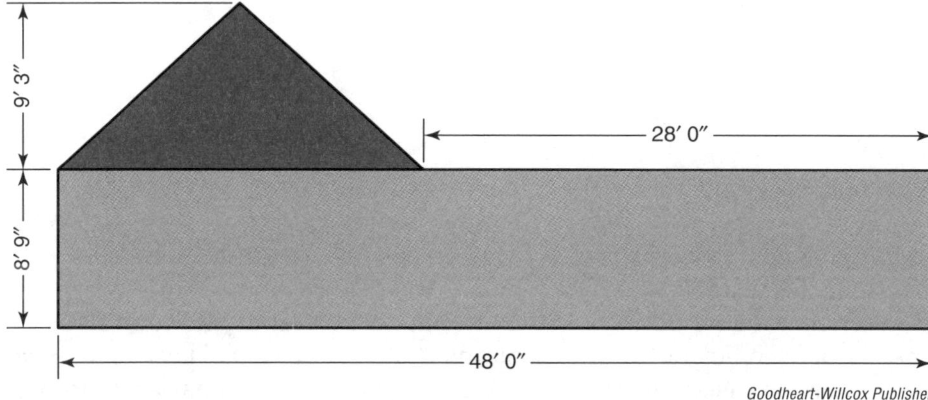

24. What is the area of the wall?

____

25. How many 4′ × 8′ pieces of sheathing are needed to cover the wall?

____

26. How many squares of siding are needed to cover the wall? (A square = 100 sq ft)

____

27. A living room measuring 15′ 0″ × 14′ 6″ with 8′ high walls needs to have the walls and ceiling painted. The paint covers 400 sq ft per gallon and the room will require two coats. How many gallons of paint will be needed? Round your answer up to the nearest gallon.

____

28. A room measuring 12′ 10″ × 14′ 6″ will have a new laminate wood floor installed. Each box of flooring contains 22.2 sq ft per box. How many boxes are needed to install the floor in this room? Round your answer up to the nearest box.

____

Name ______________________ **Date** __________ **Class** __________

29. The ceilings of a 24′ × 26′ addition and a 14′ × 12′ 8″ mudroom need to be insulated with insulation that covers 88 sq ft per bag. How many bags are needed to insulate both ceilings? Round your answer up to the nearest full bag.

____

30. Using the following image, calculate the area of the hip roof.

____

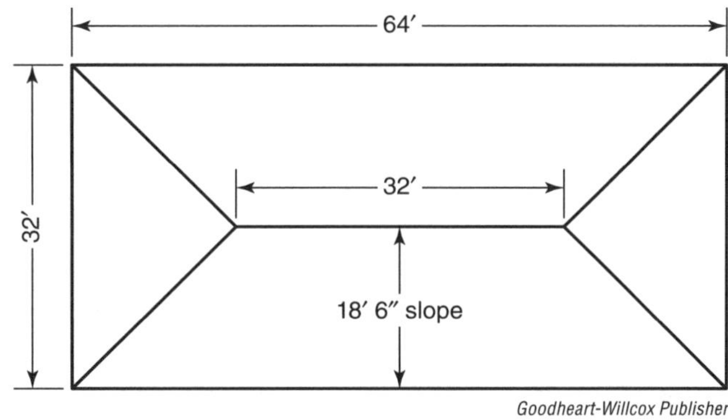

Goodheart-Willcox Publisher

Work Space/Notes

UNIT 18

Volume Measurement

Objectives

After studying this unit, you will be able to:

- Calculate the volume of a cube.
- Calculate the volume of a rectangular solid.
- Calculate the volume of irregular shaped objects.
- Convert cubic units of measure.

In the construction field, calculating volume is important for ordering materials. **Volume** is the amount of space inside a three-dimensional object consisting of length, width, and height (or in some cases thickness). The volume of an object is always measured in cubic units, such as cubic inches, cubic feet, or cubic yards. In construction, concrete is one example of a material that is always calculated in cubic yards.

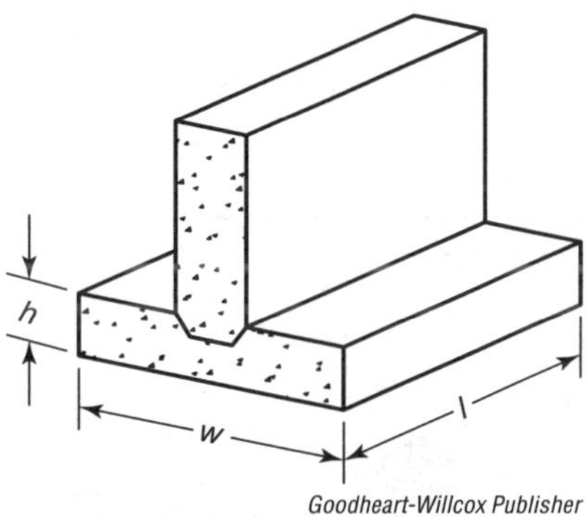

Goodheart-Willcox Publisher

Volume of a Cube

A **cube** is a three-dimensional square that has a height equal to its length and width. To determine the volume (V) of a cube, multiply three sides together and express the answer in cubic units.

$$V = l \times w \times h$$

Example 18-1

The volume of a 5″ cube is 125 cubic inches.

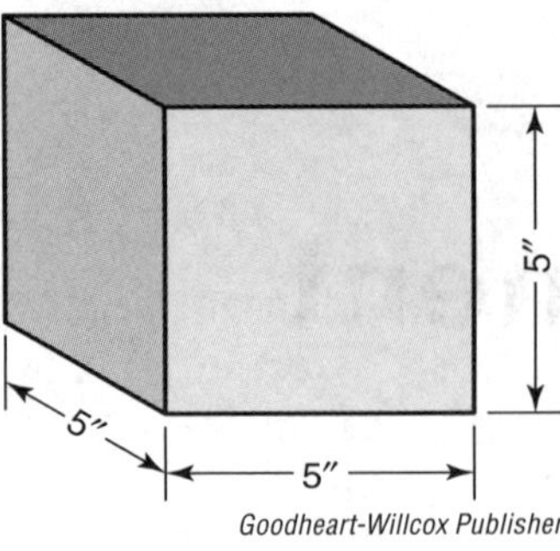

Goodheart-Willcox Publisher

$$V = s \times s \times s$$
$$V = 5'' \times 5'' \times 5''$$
$$V = 125 \text{ cu in}$$

When calculating a cube containing denominate numbers, their units combine to create a cubic unit. Volume is then expressed in cubic units using a variety of notations, such as "cu in" or "in^3."

Volume of a Rectangular Solid

A **rectangular solid** is a three-dimensional object with six sides, in which all sides are rectangles. To determine the volume of a rectangular solid, multiply the length times the width times the height of the rectangular solid and express the answer in cubic units.

$$V = l \times w \times h$$

Example 18-2

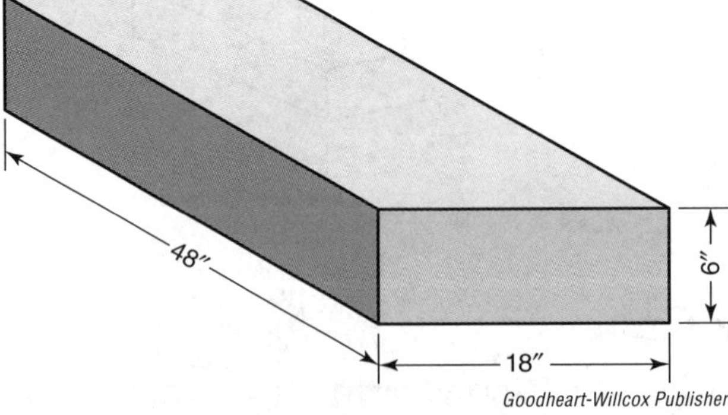

Goodheart-Willcox Publisher

The volume of the 6″ × 18″ × 48″ rectangle is 5,184 cu in.

$$V = l \times w \times h$$
$$V = 48'' \times 18'' \times 6''$$
$$V = 5{,}184 \text{ cu in}$$

Math Tip

When performing volume calculations, all measurements must be expressed in the same unit. A rectangle measuring 4″ × 8″ × 3′ could be calculated as 4″ × 8″ × 36″ = 1,152 cu in or as .333′ × .667′ × 3′ = .666 cu ft.

Volume of Irregular Shaped Objects

To determine the volume of an irregular shaped object, the shape must first be broken down into individual cubes or rectangular solids that can be calculated individually. Once the volumes of all the individual shapes have been calculated, they are then added together to determine the total volume of the irregular shaped object.

Example 18-3

To calculate the volume of the sidewalk, start by breaking the "L" into two rectangular solids.

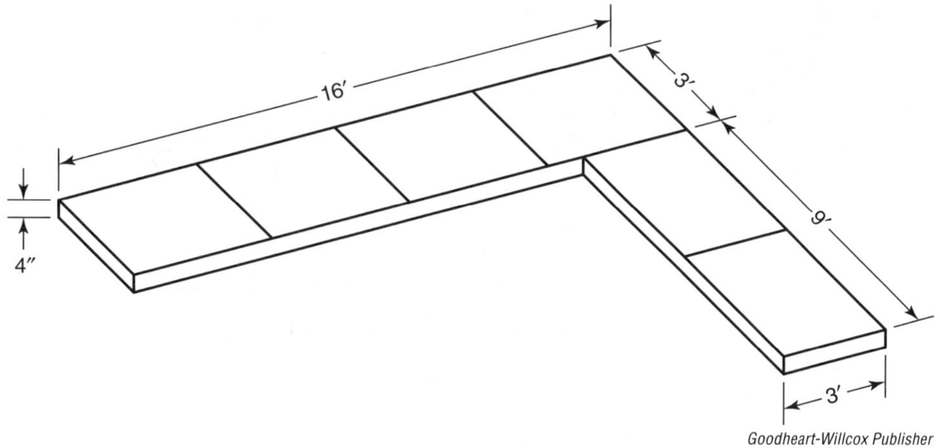

Goodheart-Willcox Publisher

Calculate the volume of the 16′ × 3′ rectangular solid.

$$V = 16' \times 3' \times 4''$$

$$V = 16' \times 3' \times 0.333'$$

$$V = 15.984 \text{ cu ft}$$

Calculate the volume of the 9′ × 3′ rectangular solid.

$$V = 9' \times 3' \times 4''$$

$$V = 9' \times 3' \times .333'$$

$$V = 8.991 \text{ cu ft}$$

Add the volumes of the two rectangular solids to determine the total volume.

$$V = 15.984 \text{ cu ft} + 8.991 \text{ cu ft}$$

$$V = 24.975 \text{ cu ft}$$

Converting Volume Measurements

In Unit 17, *Area Measurement*, the practice of converting area measurements from one square unit to another was demonstrated. In volume measure, the ability to convert measurements between cubic inches, cubic feet, and cubic yards is also a necessary skill.

In the previous example, the volume of the sidewalk was calculated to be 24.975 cu ft. Since concrete is ordered and sold by the cubic yard, the answer of 24.975 cu ft would need to be converted to cubic yards.

To make the conversions between cubic inches, feet, and yards, you must first know how many cubic inches are in a cubic foot and how many cubic feet are in a cubic yard.

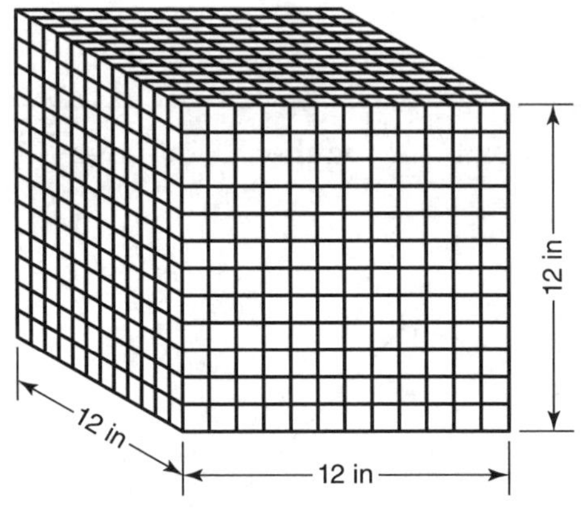

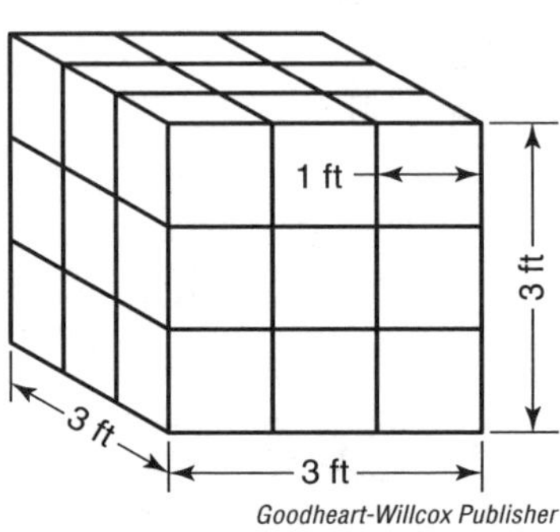

Goodheart-Willcox Publisher

A cubic foot measures 12″ × 12″ × 12″ and contains 1,728 cubic inches.

$$1 \text{ cu ft} = 12'' \times 12'' \times 12'' = 1{,}728 \text{ cu in}$$

Since there are 3′ in a yard, a cubic yard measuring 3′ × 3′ × 3′ contains 27 cubic feet.

$$1 \text{ cu yd} = 3' \times 3' \times 3' = 27 \text{ cu ft}$$

To make conversions between cubic units, the following formulas are used:

$$\text{cu in} = \text{cu ft} \times 1{,}728$$

$$\text{cu ft} = \text{cu in} \div 1{,}728$$

$$\text{cu ft} = \text{cu yd} \times 27$$

$$\text{cu yd} = \text{cu ft} \div 27$$

Example 18-4

Calculate the volume of the following concrete slab, then convert to cubic yards to determine how much concrete needs to be ordered.

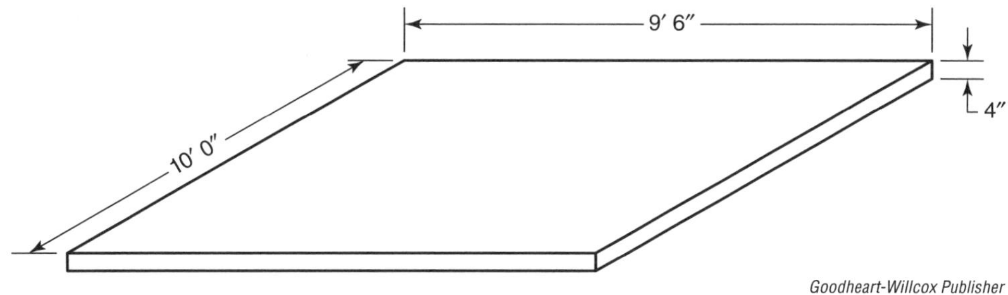

Goodheart-Willcox Publisher

$$V = 10'\ 0'' \times 9'\ 6'' \times 4''$$
$$V = 10' \times 9.5' \times .3333'$$
$$V = 31.6635 \text{ cu ft}$$

Convert the volume from cubic feet to cubic yards.

$$\text{cu ft} \div 27 = \text{cu yd}$$
$$31.6635 \text{ cu ft} \div 27 = 1.17 \text{ cu yd of concrete needed}$$

Carpentry Notes

Concrete Estimation

In the process of pouring a patio slab, first the edge forms are set. A 6″ subbase of gravel fill is placed in the edge forms before concrete can be ordered for delivery. To calculate how many cubic feet of gravel fill is needed, the volume is first calculated.

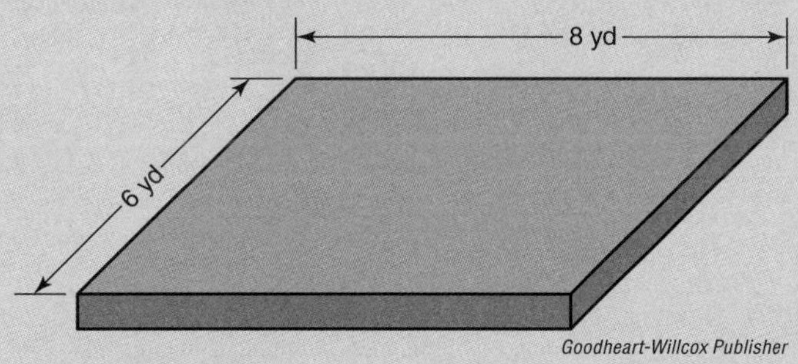

Goodheart-Willcox Publisher

$$V = l \times w \times h$$
$$V = 8 \text{ yd} \times 6 \text{ yd} \times 6''$$
$$V = 8 \text{ yd} \times 6 \text{ yd} \times 0.1666 \text{ yd}$$
$$V = 7.9968$$
$$V = 8 \text{ cu yd of gravel fill}$$

The volume is then converted to cubic feet to determine how much should be ordered.

$$\text{cu yd} \times 27 = \text{cu ft}$$
$$8 \text{ cu yd} \times 27 = 216 \text{ cu ft of gravel}$$

Work Space/Notes

Unit 18 Review

Name ______________________________ Date ____________ Class ____________

Determine the volume of the following cubes and rectangular solids, rounding all answers to two decimal places. Show all of your work.

1. Cube with 7″ sides.

   ____

2. Cube with 4′ 6″ sides.

   ____

3. Cube with 12′ 3″ sides.

   ____

4. Rectangular solid 16″ by 4″ by 2″.

____

5. Rectangular solid 48″ by 10″ by 3/4″.

____

6. Rectangular solid 5′ 3″ by 1′ 9″ by 6″.

____

7. How many cubic yards of concrete are needed to pour a sidewalk measuring 3″ thick, 3′ 6″ wide, and 48′ long?

____

Name ______________________ Date ____________ Class ____________

8. Find the number of cubic feet in a room measuring 8′ 0″ high, 11′ 8″ wide, and 13′ 9″ long.

____

9. Determine how many cubic yards of concrete are needed to pour 4 slabs, each measuring 4″ thick, 8′ 0″ wide, and 10′ 6″ long.

____

10. How many cubic yards of earth must be removed to install an in-ground swimming pool measuring 8′ 0″ × 12′ 0″ × 18′ 0″?

____

For questions 11–15, refer to the following image and calculate the cubic yards of concrete needed for the foundation wall and footer. Concrete footer sizes are generally determined by the width of the foundation wall. Footer thickness is equivalent to the width of the wall, and the footer width is 2 times the width of the wall, or a 2:1 ratio. When calculating the volumes, ignore the keyway and treat the foundation wall and foot as two straight rectangular solids.

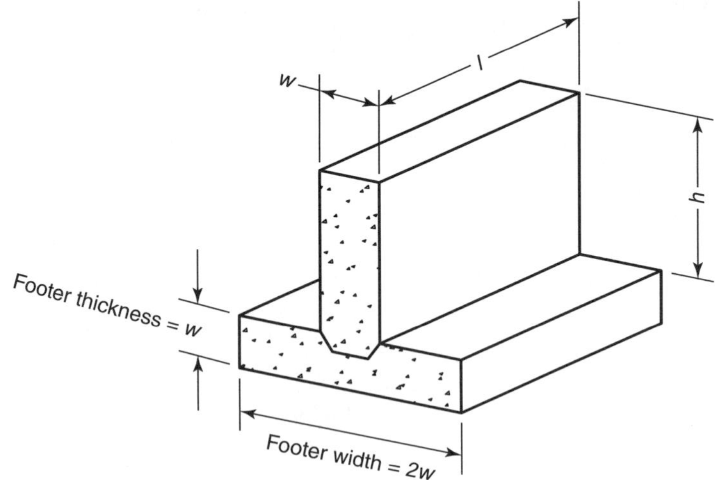

Goodheart-Willcox Publisher

	Width	Length	Height	Foundation (cubic yards)	Footer (cubic yards)
11.	8″	24′ 0″	8′ 6″		
12.	10″	18′ 9″	9′ 3″		
13.	6″	16′ 0″	6′ 9″		
14.	12″	36′ 6″	10′ 10″		
15.	8″	20′ 3″	8′ 0″		

Goodheart-Willcox Publisher

Name ______________________________ **Date** ______________ **Class** ______________

Convert the following volumes to the indicated unit.

16. 4,320 cu in = __________ cu ft

17. 9 cu yd = __________ cu ft

18. 6.5 cu ft = __________ cu in

19. 243 cu ft = __________ cu yd

20. 93,312 cu in = __________ cu yd

21. 5.5 cu yd = __________ cu in

Use the layout shown to answer questions 22–25. Calculate the volume of each area and convert to cubic yards of concrete. The concrete thickness is 4″ on the patio and all sidewalks. The driveway is 6″ thick.

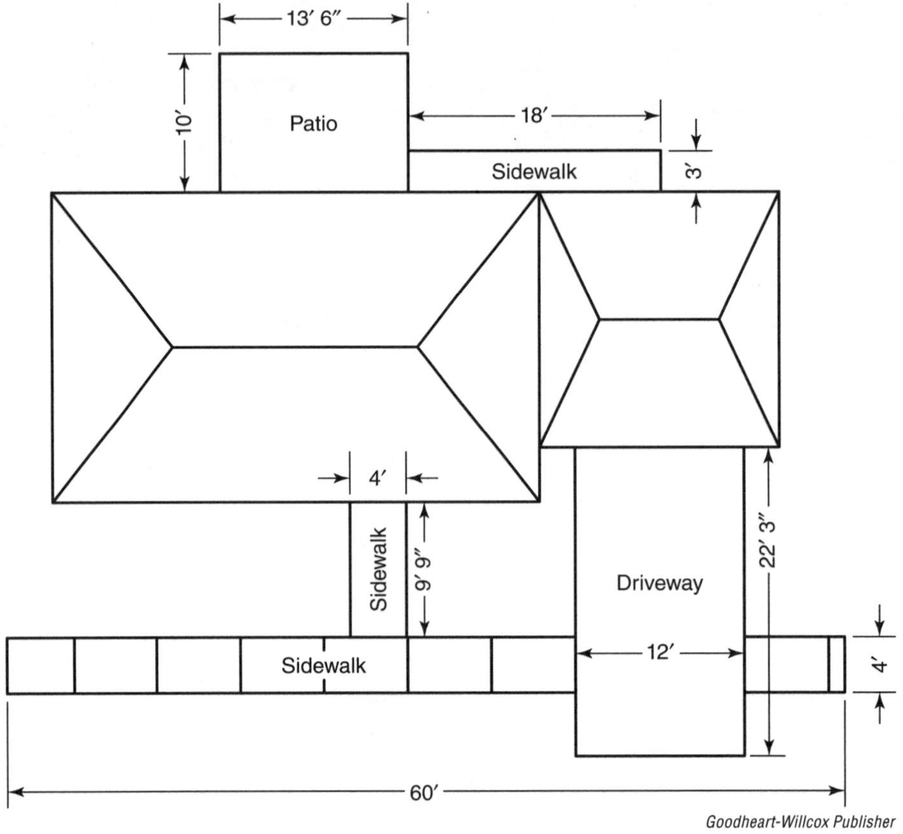

Goodheart-Willcox Publisher

22. Determine how many cubic yards of concrete are needed for the patio.

____

23. Determine how many cubic yards of concrete are needed for the sidewalk behind the house.

____

Name ______________________ Date ______________ Class ______________

24. Determine how many cubic yards of concrete are needed for the driveway.

____

25. Determine how many cubic yards of concrete are needed for all sidewalks in front of the house.

____

26. How many cubic yards of topsoil are needed to fill six planter boxes, each measuring 10′ 8″ long, 3′ 9″ wide, and 10″ deep?

____

27. How many cubic yards of earth will be removed from an excavation for a 46′ 0″ × 32′ 0″ new home? The excavation will be 6′ 6″ deep and will allow for 2′ of additional clearance around the entire foundation.

____

28. How many cubic yards of concrete are required for an L-shaped retaining wall 7′ 3″ high and 12″ thick? One side is 22′ 0″ long and the other side measures 16′ 0″.

____

29. A 4″ base of gravel fill is needed to grade the basement of a new home before the concrete floor is poured. The dimensions of the basement are 46′ 8″ × 31′ 8″. How many cubic yards of gravel are needed?

____

30. Determine how many cubic yards of concrete will need to be ordered for the 4″ concrete slab shown below.

____

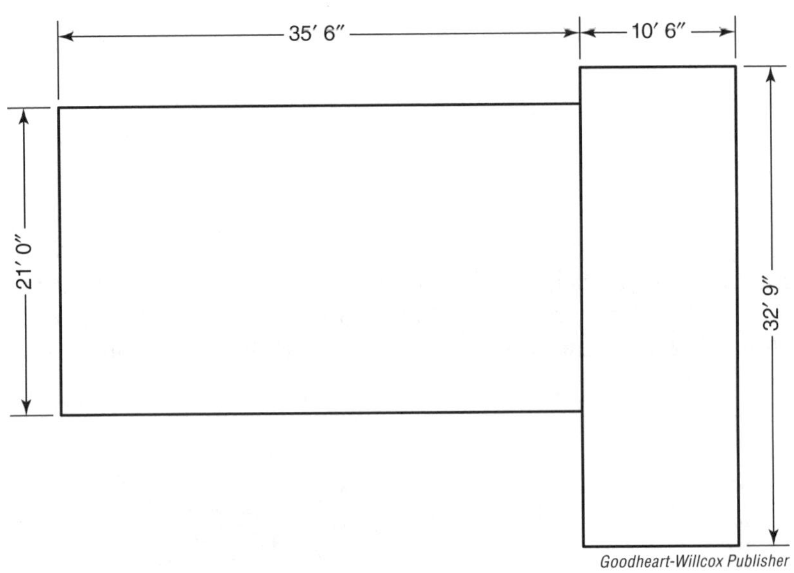

Goodheart-Willcox Publisher

UNIT 19

Board Foot Measurement

Objectives

After studying this unit, you will be able to:

- Determine nominal and dressed sizes of lumber.
- Calculate board foot quantities.

In the construction industry, not all materials are measured and calculated alike. When it comes to estimating materials for a project or a quote, a skilled carpenter must know how to calculate a wide variety of products accurately. For example, sheet goods are calculated by area and are sold by the sheet. Concrete is calculated by its volume and is sold by the cubic yard. Millwork and trims are sold by the lineal foot. Rough lumber purchased at a mill and random sized boards are calculated by the board foot.

Another factor that must be considered when calculating lumber needs is the difference between the nominal and dressed sizes of lumber. If a project requires a board that actually measures 1 1/2″ × 10″, then a 2 × 12 board will need to be purchased.

Nominal and Dressed Sizes of Lumber

When trees are ready to harvest, they are cut by loggers, loaded onto trucks, and taken to a mill to be cut into dimensional lumber. During the first stage of the milling process, the logs are analyzed and then milled to get the best yield from each log.

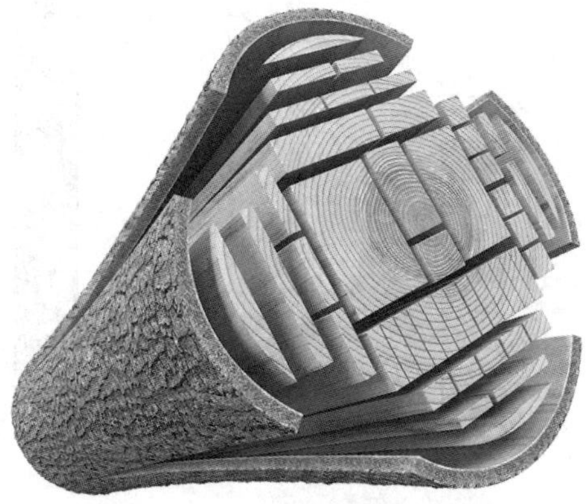

Slavoljub Pantelic/Shutterstock.com

This first cut reduces the boards to their **nominal size**, the size from which the boards are named. These boards are considered **rough cut** and contain a tremendous amount of moisture. Depending on the species and quality of the wood, they are stacked with stickers between each row of boards, and then are either air-dried or kiln-dried to reduce the moisture to an acceptable level.

Once dried, the lumber is then planed smooth and reduced in thickness and width to its surfaced size, or **dressed size**, the size the finished board actually measures.

Nominal Size	Dressed Size	Nominal Size	Dressed Size
1 × 2	3/4″ × 1 1/2″	2 × 8	1 1/2″ × 7 1/4″
1 × 3	3/4″ × 2 1/2″	2 × 10	1 1/2″ × 9 1/4″
1 × 4	3/4″ × 3 1/2″	2 × 12	1 1/2″ × 11 1/4″
1 × 6	3/4″ × 5 1/2″	4 × 4	3 1/2″ × 3 1/2″
1 × 8	3/4″ × 7 1/4″	4 × 6	3 1/2″ × 5 1/2″
1 × 10	3/4″ × 9 1/4″	4 × 8	3 1/2″ × 7 1/4″
1 × 12	3/4″ × 11 1/4″	4 × 10	3 1/2″ × 9 1/4″
2 × 2	1 1/2″ × 1 1/2″	4 × 12	3 1/2″ × 11 1/4″
2 × 3	1 1/2″ × 2 1/2″	6 × 6	5 1/2″ × 5 1/2″
2 × 4	1 1/2″ × 3 1/2″	6 × 8	5 1/2″ × 7 1/4″
2 × 6	1 1/2″ × 5 1/2″		

Goodheart-Willcox Publisher

Four simple patterns emerge in the sizing of dressed lumber making it easy to remember all the sizes.

- The thickness of all 1 × material is 3/4″.
- The thickness of all 2 × material is 1 1/2″.
- The width of all boards 6″ or less is 1/2″ smaller than its nominal size.
- The width of all boards 8″ or greater is 3/4″ smaller than its nominal size.

Lumber is always referred to by its nominal size (2 × 8) even though its dressed size is smaller (1 1/2″ × 7 1/4″).

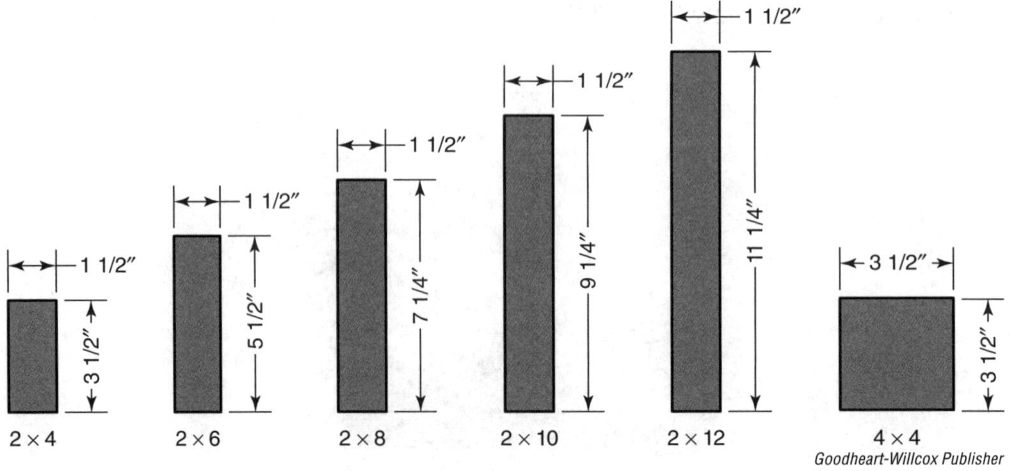

Goodheart-Willcox Publisher

Lumber can also be sold in its rough cut state. The boards are cut and dried but not planed to a specific dressed size. Since the thickness of this rough stock is only approximate, the lumber is then referred to using the "quarter" sizing method. Using this method, a 4/4 board describes a board approximately 1″ thick and a 5/4 board describes a board approximately 1 1/4″ thick. The quarter designation is simply a fraction that has not been reduced.

- 4/4 = 1″ thick
- 5/4 = 1 1/4″ thick
- 6/4 = 1 1/2″ thick
- 8/4 = 2″ thick

Calculating Board Foot

When rough cut boards are purchased from a mill, or when lumber is purchased in random lengths and widths, they cannot be sold by the piece or lineal foot because there is no consistency in their dimensions. When this occurs, lumber quantity is then determined by the **board foot**, which calculates the volume within the lumber. A board foot of lumber contains 144 cubic inches of lumber and is equivalent to a piece of stock measuring 1″ thick, 12″ wide, 1′ long.

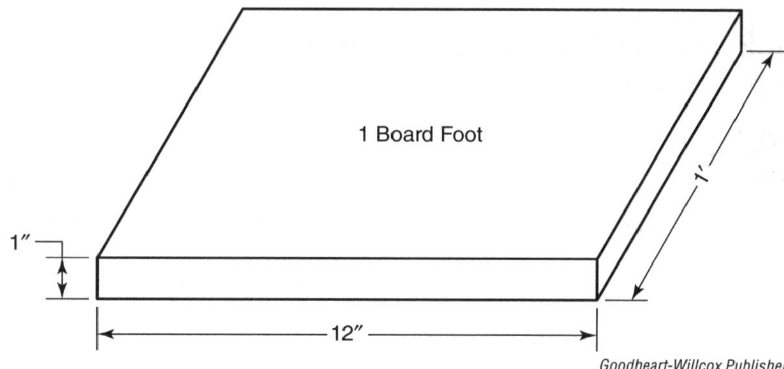

Goodheart-Willcox Publisher

To calculate lumber by the board foot, a standard formula uses the board's thickness (t), width (w), and length (l) to determine the quantity.

$$\text{Board foot} = \frac{t\ (\text{inches}) \times w\ (\text{inches}) \times l\ (\text{feet})}{12}$$

When a board needs to be calculated, insert the board's dimensions into the formula.

Math Tip

When calculating board feet, the nominal dimensions of lumber are always used.

Example 19-1

How many board feet are in a 2 × 10 board measuring 14′ long?

$$\frac{2'' \times 10'' \times 14'}{12} = \frac{280}{12} = 23.333 \text{ bd ft}$$

Since a board foot is a measurement of a board's volume, rather than specific dimensions, materials of random sizes can still be equivalent to a board foot.

1″ × 3″ × 4′ $\frac{1'' \times 3'' \times 4'}{12} = 1 \text{ bd ft}$

2″ × 2″ × 3′ $\frac{2'' \times 2'' \times 3'}{12} = 1 \text{ bd ft}$

1″ × 6″ × 2′ $\frac{1'' \times 6'' \times 2'}{12} = 1 \text{ bd ft}$

Goodheart-Willcox Publisher

When calculating the board feet of multiple boards of the same size, the variable (quantity of pieces) is added to the formula.

Example 19-2

Calculate the board feet in 16 pieces of 1 × 8 boards measuring 8′ long.

$$\frac{16 \times 1'' \times 8'' \times 8'}{12} = \frac{1024}{12} = 85.333 \text{ bd ft}$$

Another board foot formula can be used when calculating boards whose lengths do not measure an even foot.

$$\text{Board foot} = \frac{t \text{ (inches)} \times w \text{ (inches)} \times l \text{ (inches)}}{144}$$

In this formula, the thickness, width, and length are calculated in inches. Since there are 144 square inches in a square foot, then the answer is divided by 144.

Example 19-3

How many board feet are in a 2 × 6 board measuring 13′ 5″ long?

Convert the length to inches: 13 × 12 = 156″ + 5″ = 161″

$$\frac{2'' \times 6'' \times 161''}{144} = \frac{1932}{144} = 13.41667 \text{ bd ft}$$

Carpentry Notes

Calculating Board Foot

A woodworker wants to build the standing shelf unit shown below. Because of the species of material needed for the project, she will purchase the material as rough cut lumber and plane it down to dressed size in her workshop.

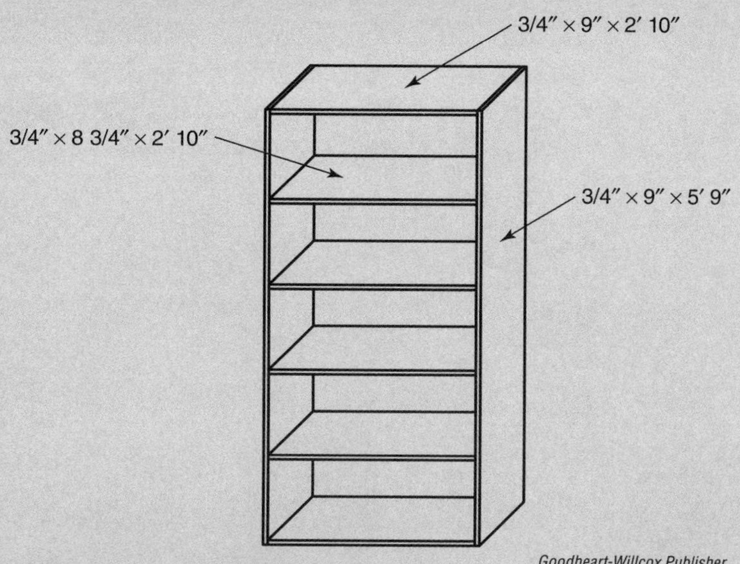

Goodheart-Willcox Publisher

She determines the measurements needed for each part of the unit, then decides what size board needs to be purchased and planed to fit. Because rough cut material is being purchased and then planed, all material is rounded up to whole inches or feet. She calculates the board feet of the pieces individually using the rounded-up sizes and adds up the totals of each calculation to determine the total board feet of material needed from the mill. The back of the shelf is 1/4″ plywood and is not calculated using board feet.

Part	Quantity	Part Size Needed	Rough Cut Material Estimate		Board Feet
Sides	2	3/4″ × 9″ × 5′ 9″	$\frac{2 \times 1'' \times 10'' \times 6'}{12}$	=	10 bd ft
Ends	2	3/4″ × 9″ × 2′ 10″	$\frac{2 \times 1'' \times 10'' \times 3'}{12}$	=	5 bd ft
Shelves	4	3/4″ × 8 3/4″ × 2′ 10″	$\frac{4 \times 1'' \times 9'' \times 3'}{12}$	=	9 bd ft

10 bd ft + 5 bd ft + 9 bd ft = 24 bd ft of material

Work Space/Notes

Unit 19 Review

Name ______________________ Date ____________ Class ____________

Solve the following problems relating to dressed lumber sizes.

1. What is the width of a 2 × 4 and a 2 × 6 laid side by side?

   ____

2. What is the thickness of a 1 × 8 and a 2 × 8 stacked together?

   ____

3. What is the width of a 2 × 10 and a 2 × 12 laid side by side?

   ____

4. How thick is a 4 layer built-up beam made from 2 × 12 stock?

   ____

Use the following beam to answer questions 5–6.

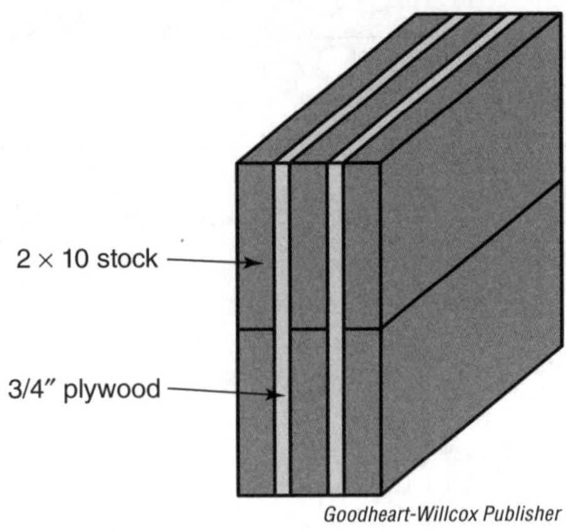

Goodheart-Willcox Publisher

5. How thick is the built-up beam?

____

6. How tall is the built-up beam?

____

7. How many rows of 2 × 4s can be stacked in a 35″ tall storage rack?

____

Name ______________________ **Date** __________ **Class** __________

8. How thick is an exterior 2 × 6 wall with 7/16″ exterior sheathing and 1/2″ sheetrock?

____

Use the following board to answer questions 9–10.

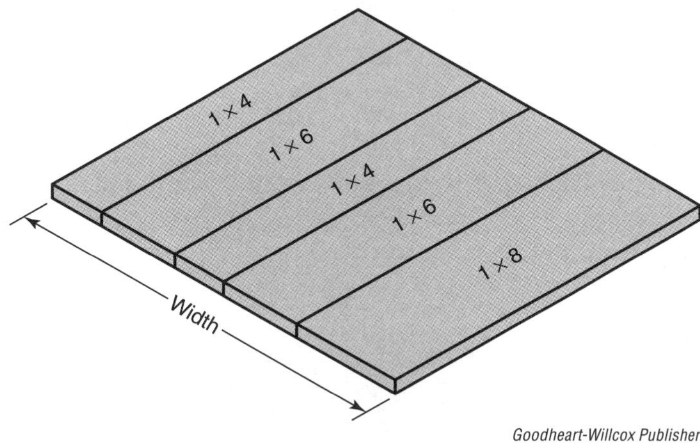

Goodheart-Willcox Publisher

9. What is the nominal width of the glued-up board?

____

10. What is the actual width of the glued-up board?

____

Solve the following board foot problems using nominal sizes. Round all answers to three decimal places. Show all of your work.

11. How many board feet are in 8 pieces of $1 \times 10 \times 14'$ lumber?

____

12. How many board feet are in 15 pieces of $2 \times 8 \times 10'$ lumber?

____

13. How many board feet are in 22 pieces of $1 \times 8 \times 16'$ lumber?

____

14. How many board feet are in 35 pieces of $2 \times 6 \times 12'$ lumber?

____

15. How many board feet are in 44 pieces of $1 \times 12 \times 14'$ lumber?

____

Name ______________________ **Date** ____________ **Class** ____________

16. How many board feet are in 6 pieces of 2 × 10 × 12′ lumber?

____

17. How many board feet are in 22 pieces of 2 × 12 × 16′ lumber?

____

The floor system was built with rough cut lumber. Use the image to answer questions 18–22.

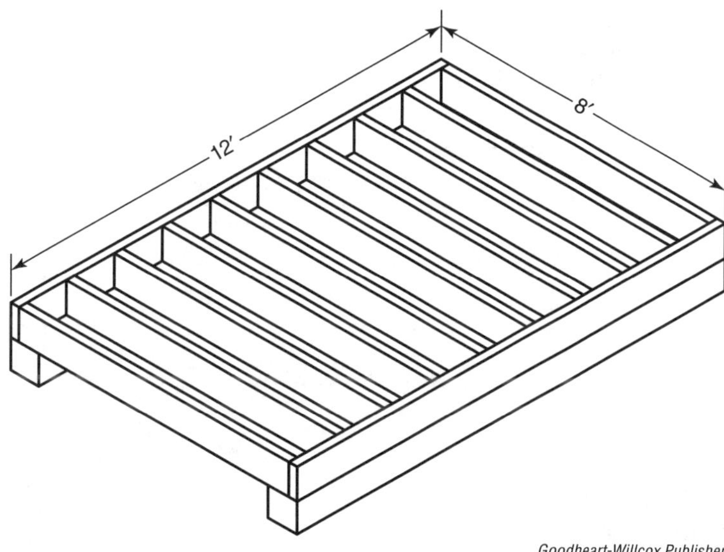

Goodheart-Willcox Publisher

18. Calculate the board feet of the two 6 × 6 pressure-treated runners beneath the floor system.

____

19. The floor frame was built using 2 × 6 stock. Determine the board feet for the 10 floor joists.

____

20. Determine the board feet of the two 12′ rim joists.

____

21. Rough cut 1″ floorboards will be used as sheathing for the shed floor. How many board feet of planks need to be purchased?

____

22. If the floor is sheathed with pressure-treated 2 × 6 × 12′ boards, how many boards would be needed?

____

Name ______________________ Date __________ Class __________

23. A woodworker purchased rough cut walnut boards from a mill for a project. Determine from the following list the total number of board feet of walnut boards purchased.

    (12) 4/4″ × 6″ × 10′ 0″ boards
    (4) 4/4″ × 10″ × 14′ 0″ boards
    (4) 4/4″ × 10″ × 8′ 0″ boards
    (2) 4/4″ × 14″ × 12′ 0″ boards

    ____

24. A woodworker purchased 18 white pine boards measuring 4/4″ × 8″ × 14′, and 13 red oak boards measuring 4/4″ × 6″ × 10′ from a mill. What is the total board feet of each species purchased?

    ____

25. Currently rough white pine is sold at $0.88 per board foot and red oak sells for $1.92 per board foot. What was the total cost of the lumber the woodworker purchased?

    ____

Work Space/Notes

Section 4 Exam

Name ______________________________ Date ____________ Class ____________

Round all answers to three decimal places as needed.

1. Convert 4′ 7″ to inches.

   ____

2. Convert 79 1/2″ to feet and inches.

   ____

3. Convert 118 1/4″ to feet and inches.

   ____

4. Convert 54″ to decimal equivalents in feet.

   ____

5. Convert 18.8125′ to fractional feet and inches.

____

6. What is the perimeter of a building that measures 28′ wide and 36′ long?

____

7. What is the perimeter of a house that measures 34′ 6″ wide and 56′ long?

____

8. Determine the area of a rectangle with a length of 9′ 8″ and a width of 5′ 6″.

____

9. Determine the area of a triangle with a 24′ base and an 11′ 6″ height.

____

Name ______________________________ Date ____________ Class ____________

10. How many 4′ x 8′ sheets of subfloor are needed to cover the floor joists of a 26′ × 46′ 6″ two-story home?

____

11. Convert 648 sq in to sq ft.

____

12. Convert 7.5 sq ft to sq in.

____

13. Determine the volume of a rectangular solid 18″ × 5″ × 3.5″.

____

14. Using dressed sizes, what is the total width of a 2 × 4, 2 × 6, 2 × 8, and a 2 × 10 laid side by side?

____

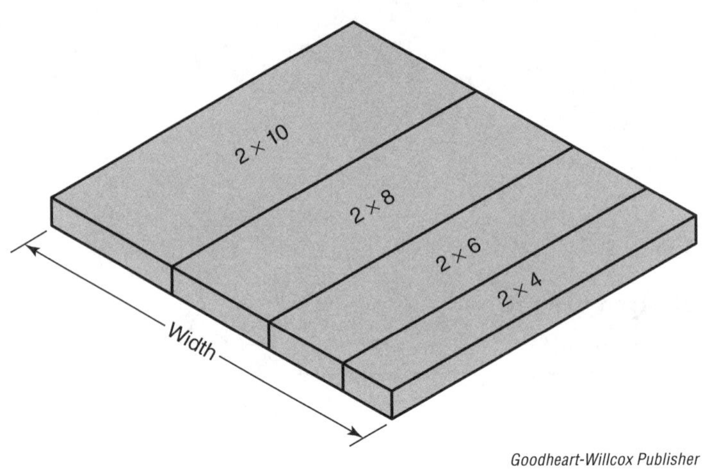

Goodheart-Willcox Publisher

15. Using nominal sizes, a stack of 36 pieces of 2 × 8 × 12′ boards consists of how many board feet?

____

16. Determine the volume of a rectangular solid measuring 12.5″ × 5″ × 3.5″.

____

Name ______________________ Date ____________ Class ____________

17. Convert 24 cu yd to cu ft.

____

18. Convert 486 cu ft to cu yd.

____

19. How many 16′ sill plates are needed for a 28′ × 46′ foundation wall?

____

20. How many 10′ pieces of aluminum drip edge are needed around the perimeter of a 44′ long gable roof with a 13′ 6″ slope?

____

21. What is the perimeter of a room measuring 18′ 3″ long and 138″ wide?

____

22. Determine how many cubic yards of concrete are needed to pour 2 slabs, each measuring 4″ thick, 12′ 0″ wide, and 18′ 3″ long.

____

23. How many board feet are in 18 pieces of lumber measuring 2 × 6 × 14′?

____

24. A carpenter will be using 1″ thick plank floorboards to sheath a barn floor measuring 44′ × 62′. How many board feet of planks will need to be purchased?

____

25. A 9′ 3″ × 17′ 6″ wall will be covered with 3/4″ tongue-and-groove knotty pine. How many board feet are needed for the job?

____

SECTION 5

Percentages and Right Angles

Key Terms

3-4-5 method
hypotenuse
percent symbol (%)
percentage
Pythagorean theorem
right angle
right triangle
square
square root

UNIT 20

Percentages

Objectives

After studying this unit, you will be able to:

- Convert a percentage to a fraction.
- Convert a fraction to a percentage.
- Convert a percentage to a decimal.
- Convert a decimal to a percentage.
- Demonstrate the ability to calculate a percentage.

Percentages are useful when comparing a quantity to a whole. A **percentage** is a part of a whole, representing a fraction of 100. For example, if 65 out of 100 boards have been cut, then it is said that 65% of the boards have been cut. The **percent symbol (%)** identifies a percentage. Percentages are commonly used in the industry to express:

- The amount of moisture in lumber.
- Salaries, taxes, and deductions.
- Equipment or materials on sale.
- Waste factors on supplies and materials.
- Interest rates.

The ability to solve problems involving percentages is essential in the construction trades. Because percentages are a representation, rather than an actual number, they must be converted to either a fraction or a decimal before being used in a math operation.

Converting a Percentage to a Fraction

There are times when a percentage must be converted to a fraction to determine the exact quantity it represents. Since any number identified with a percent symbol already refers to that number out of 100, the simplest way to convert the number to a fraction is to remove the percent symbol and place the number as a numerator over a denominator of 100. Then reduce the fraction to lowest terms. For example, 50% means 50 out of 100.

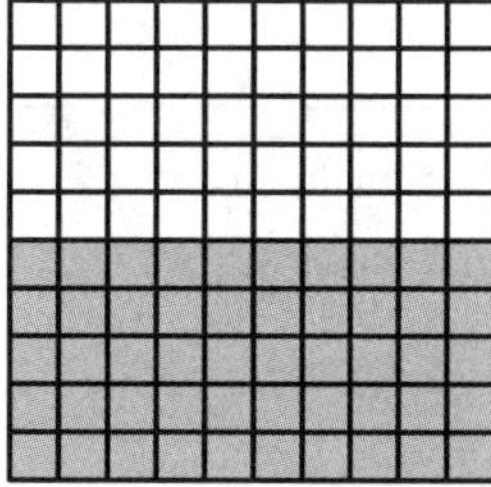

Goodheart-Willcox Publisher

Math Tip

The percent symbol always identifies a number as parts per 100.

Example 20-1

Any percentage can easily be converted to a fraction.

$$45\% = \frac{45}{100} = \frac{9}{20}$$

$$88\% = \frac{88}{100} = \frac{22}{25}$$

$$200\% = \frac{200}{100} = 2$$

If the percentage is a fraction, remove the percent symbol and multiply the fraction by 1/100. If the percentage is a mixed number, first convert it to an improper fraction, remove the percent symbol, and then multiply the denominator by 1/100.

Example 20-2

$$\frac{3}{4}\% = \frac{3}{4} \times \frac{1}{100} = \frac{3}{400}$$

$$15\frac{1}{3}\% = \frac{46}{3}\% = \frac{46}{3} \times \frac{1}{100} = \frac{46}{300} = \frac{23}{150}$$

$$125\frac{1}{2}\% = \frac{251}{2}\% = \frac{251}{2} \times \frac{1}{100} = \frac{251}{200} = 1\frac{51}{200}$$

Converting a Percentage to a Decimal

When it is necessary to convert a percentage to a decimal, remove the percent sign and divide by 100. This division can be done simply by moving the decimal point two place values to the left.

Math Tip

If the number being changed from a percent to a decimal is only one digit, then a zero must be added before the number in order to move the decimal point two places to the left.

Example 20-3

Any percentage can easily be converted to a decimal.

$$19.75\% = 0.1975$$

$$7\% = 0.07$$

$$122\% = 1.22$$

$$215\% = 2.15$$

Converting a Decimal to a Percentage

The process of converting a decimal to a percentage is simply the opposite of what was done when converting a percentage to a decimal. The decimal point is moved two places to the right and the percent sign is added.

Example 20-4

With 0.75 of the project completed, what is the percentage of work that has been done so far?

$$0.75 = 75\%$$

75% of the project is complete.

Example 20-5

Any decimal can easily be converted to a percentage.

$$0.35 = 35\%$$

$$0.02 = 2\%$$

$$2.25 = 225\%$$

Converting a Fraction to a Percent

There are times when a fraction may need to be converted to express its value as a percentage. To convert a fraction or a mixed number to a percentage, it must first be converted to a decimal. It is then converted to a percentage by moving the decimal two places to the right and adding the percent symbol.

Example 20-6

If 3 out of 8 workers call in sick, what percentage of the workforce is out sick?

$$\frac{3}{8} = 0.375 = 37.5\%$$

37.5% of the workforce is out sick.

Example 20-7

Any fraction can easily be converted to a percentage.

$$\frac{13}{16} = 0.8125 = 81.25\%$$

$$1\frac{5}{8} = 1.625 = 162.5\%$$

$$3\frac{1}{2} = 3.50 = 350\%$$

Calculating Percentages

In order to solve percent problems, you must identify the three basic parts of the problem and then determine the correct operation to solve the problem. The three parts of any percent problem contain the whole (base), the percentage, and the amount (part). You must first identify which two of the three basic parts you have and then use the correct operation to solve the missing part.

Missing Piece	Operation
Amount	Percentage × Whole
Percentage	Amount ÷ Whole
Whole	Amount ÷ Percentage

Goodheart-Willcox Publisher

Example 20-8

What is 12% of 275? The *percentage* and *whole* are known, the *amount* is missing.

$$\begin{aligned} \text{Amount} &= \text{Percentage} \times \text{Whole} \\ &= 12\% \times 275 \\ &= 0.12 \times 275 \\ &= 33 \end{aligned}$$

12% of 275 is 33.

Example 20-9

16 is what percent of 88? The *amount* and the *whole* are known, the *percentage* is missing.

$$\begin{aligned} \text{Percentage} &= \text{Amount} \div \text{Whole} \\ &= 16 \div 88 \\ &= 0.181818 \\ &= 18.1818\% \end{aligned}$$

16 is 18.1818% of 88.

Example 20-10

42 is 25% of what number? The *amount* and the *percentage* are known, the *whole* is missing.

$$\begin{aligned} \text{Whole} &= \text{Amount} \div \text{Percentage} \\ &= 42 \div 0.25 \\ &= 168 \end{aligned}$$

42 is 25% of 168.

Carpentry Notes

Profit Percentages

When estimating costs for any type of job, it is important to know how much money is being figured for materials, labor, and equipment, and what percentage will become your profit. A home addition project that has been bid at $45,000 has the following estimated costs:

Materials—$29,000

Labor—$4,725

Equipment—$1,300

To determine what percentage of the overall cost will be profit for the builder, start by adding the cost of materials, labor, and equipment.

$$\$29,000 + \$4,725 + \$1,300 = \$35,025$$

Subtract the total cost of materials, labor, and equipment from the overall bid for the project to determine the amount of profit for the builder.

$$\$45,000 - \$35,025 = \$9,975$$

Now that the amount of profit has been determined, divide by the whole to determine the percentage.

$$\begin{aligned} \text{Percentage} &= \text{Amount} \div \text{Whole} \\ &= \$9,975 \div \$45,000 \\ &= 0.2217 \\ &= 22.17\% \end{aligned}$$

22.17% of the estimate is profit.

Unit 20 Review

Name ______________________ Date ____________ Class ____________

Express the following percentages as fractions reduced to lowest terms. Show all of your work.

1. 73%

 ____

2. 1%

 ____

3. 165%

 ____

4. $\frac{2}{7}\%$

 ____

5. $\frac{5}{16}\%$

 ____

6. $1\frac{1}{8}\%$

 ____

7. $32\frac{2}{3}\%$

____

8. $13\frac{1}{6}\%$

____

Express the following percentages as a decimal. Show all of your work.

9. 23%

____

10. 98%

____

11. 24.5%

____

12. 18.35%

____

Name ______________________________ **Date** ____________ **Class** ____________

13. 0.138%

____

14. 0.028%

____

15. 319%

____

16. $\frac{1}{5}\%$

____

Express the following decimals as percentages. Show all of your work.

17. 0.73

____

18. 0.14

____

19. 0.046

____

20. 0.0062

____

21. 4.94

____

22. 2.06

____

23. 96.02

____

24. 1.0006

____

Name ______________________ **Date** ____________ **Class** ____________

Express the following fractions as percentages. Show all of your work.

25. $\frac{7}{10}$

____

26. $\frac{13}{64}$

____

27. $1\frac{5}{8}$

____

28. $2\frac{1}{5}$

____

29. $4\frac{3}{20}$

____

30. $3\frac{43}{100}$

____

Calculate the following problems rounding to two decimal places when needed. Show all of your work.

31. What is 8% of 150?

____

32. What is 27% of 320?

____

33. 9 is what percent of 81?

____

34. 32 is what percent of 160?

____

35. 12 is 40% of what number?

____

36. 54 is 96% of what number?

____

Name ______________________ Date ____________ Class ____________

37. A 1,530 sq ft home had a 345 sq ft addition constructed. What percentage of the completed home can be attributed to the addition?

____

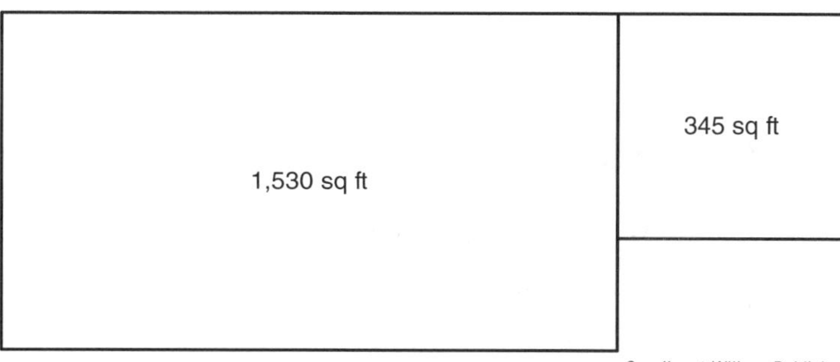

Goodheart-Willcox Publisher

38. The area of the exterior walls of a house is 1,969 sq ft. The contractor for the project ordered a total of 2,200 sq ft of sheathing for the job, including extra for rough openings and waste. What percentage was figured in for rough openings and waste?

____

39. The total cost of construction for a home was $197,000. The roofing and siding will be installed by a subcontractor for 14% of the total costs. How much is the subcontractor being paid for the roofing and siding?

____

40. The following plot plan shows a home located on a lot. What percentage of the lot does the home occupy?

____

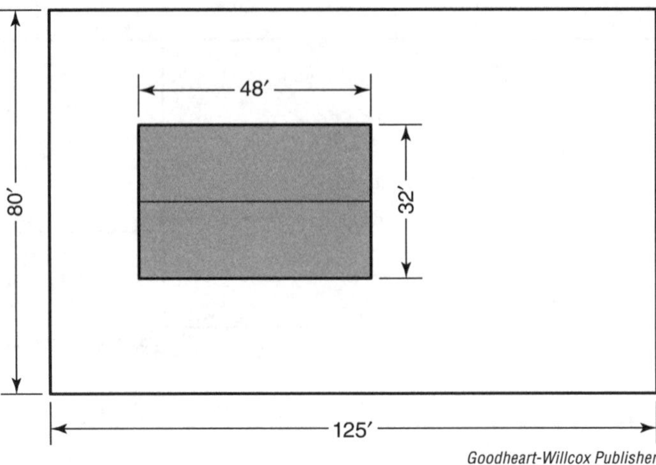

Goodheart-Willcox Publisher

UNIT 21

Right Angles

Objectives

After studying this unit, you will be able to:

- Demonstrate an understanding of right triangle theory.
- Establish a right angle using the 3-4-5 method.
- Determine the hypotenuse of a right triangle using the Pythagorean theorem.

There are many types of triangles, but the one most commonly used in construction is the right triangle. While every triangle is a three-sided shape, a **right triangle** always contains a 90° angle. A 90° angle is formed by two perpendicular lines and is also known as a **right angle**.

Many tasks in the construction field require constructing right angles, such as staking out a building or building a wall that is square. The right triangle is also critical in a variety of construction tasks including determining the length of rafters and stair stringers. Knowing the mathematical principles of right triangles is essential for constructing sound and stable structures.

Math Tip

Only a triangle containing a right angle is a right triangle.

Right Triangle Theory

A 90° angle is often illustrated with a square box in the corner. The two sides of the triangle that extend from the square corner are identified by the lowercase letters a and b. Across from the square corner is the longest side of the triangle, called the **hypotenuse**, identified by the letter c.

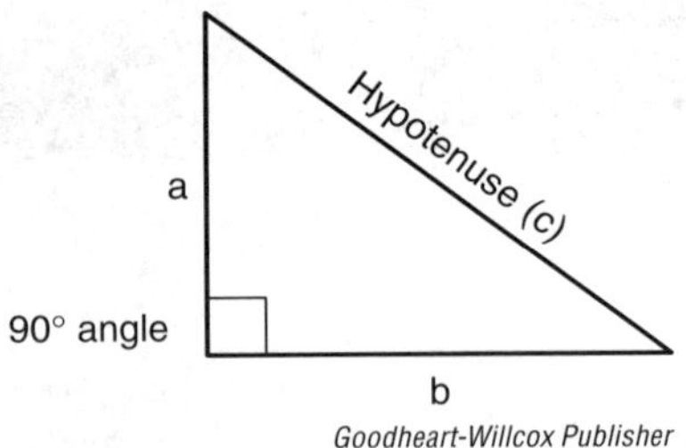

Goodheart-Willcox Publisher

A triangle is only a right triangle if the length of the hypotenuse is in specific proportions with the lengths of the other two sides. Two methods are used to determine these lengths, the 3-4-5 method and the Pythagorean theorem.

Example 21-1

To determine if a wall is square prior to applying exterior sheathing, two sides of the wall are used to make a right triangle. The lengths of the sides determine what length the hypotenuse needs to be to form a right triangle.

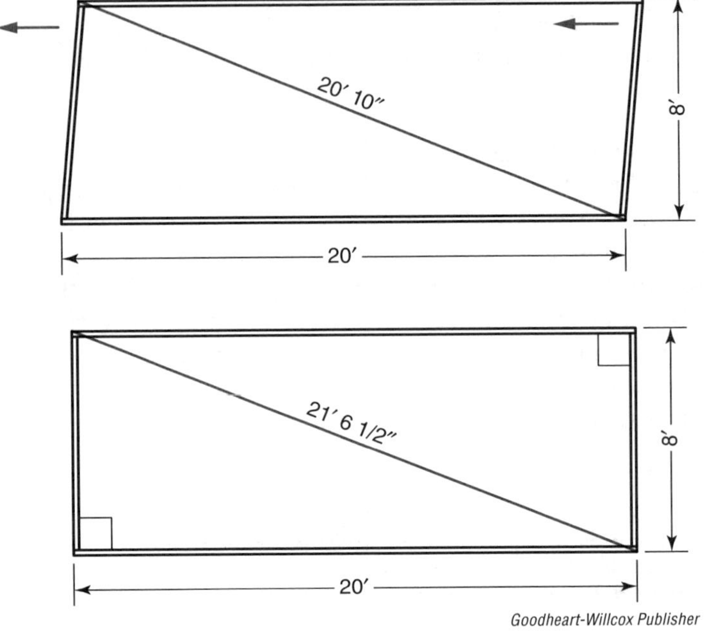

Goodheart-Willcox Publisher

A wall measuring 8′ high and 20′ long has a diagonal measurement that currently measures 20′ 10″. The hypotenuse should measure 21′ 6 1/2″ for the wall to produce a 90° corner or square wall. To ensure a square wall, the hypotenuse should be lengthened to measure 21′ 6 1/2″ by adjusting the walls. The wall leans to the right and must be shifted slightly to the left to increase the diagonal measurement until the proper hypotenuse is achieved.

Once the wall is adjusted and the diagonal length equals the hypotenuse, the wall is then considered square. This method of squaring the wall works on multiple construction tasks when a square corner is essential.

The 3-4-5 Method

A simple method used to establish a right angle in a triangle is the **3-4-5 method**. The method states that if a triangle is laid out using 3, 4, and 5 units as the lengths of the sides, the triangle is a right triangle. The hypotenuse will measure 5 units.

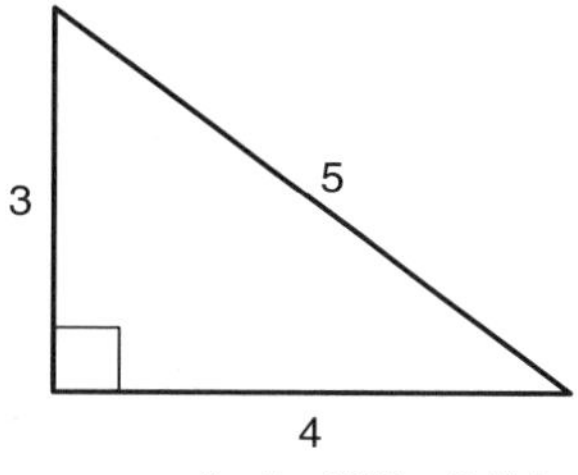

Goodheart-Willcox Publisher

Any multiples of 3-4-5 will work to establish a right angle, such as 6-8-10 or 12-16-20. For accuracy, it is best to use a multiple that best fits the size of the project. Different sized layouts can be squared using variations of the 3-4-5 method.

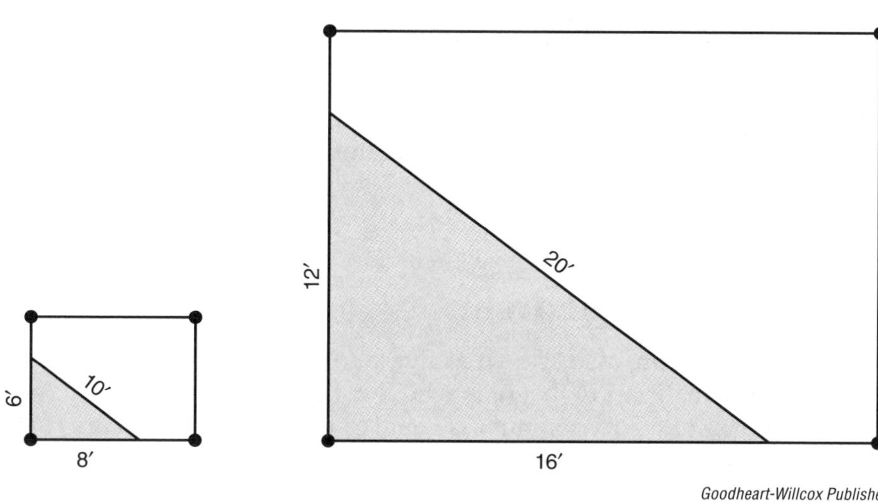

Goodheart-Willcox Publisher

Example 21-2

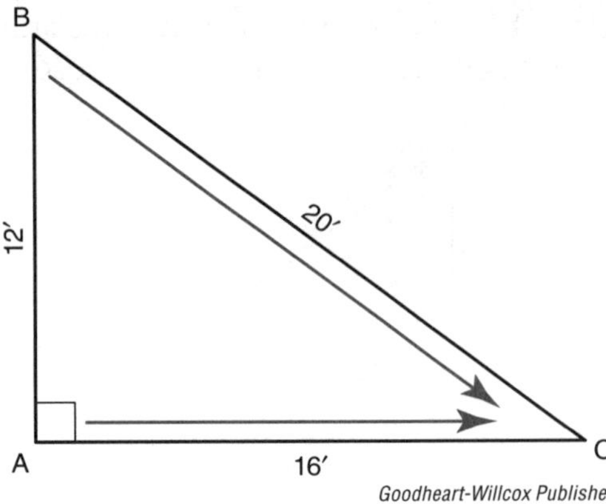

Goodheart-Willcox Publisher

A 12′ × 16′ patio is to be staked out. The first corner stake is established at A. Then a stake is placed 12′ away at B. To make sure the patio is square, the placement of the third stake at C is located by viewing the two sides as part of a triangle and determining the hypotenuse. Using the 3-4-5 method, a right triangle with sides of 12′ and 16′ has a hypotenuse that measures 20′.

The third stake is placed by measuring simultaneously with two tapes 16′ and 20′. The third stake is placed where the two tapes intersect at C.

The Pythagorean Theorem

The **Pythagorean theorem**, also known as Pythagoras's theorem, is named after the ancient Greek mathematician Pythagoras who lived over 2,700 years ago. The theory states that the square of the hypotenuse is equal to the sum of the squares of the two shorter sides, $a^2 + b^2 = c^2$. **Square** is defined as a number multiplied by itself, such as $4^2 = 16$ or $4 \times 4 = 16$.

The Pythagorean theorem can be demonstrated using the 3-4-5 variables.

$$a^2 + b^2 = c^2$$

$$3^2 + 4^2 = 5^2$$

$$9 + 16 = 25$$

The formula can be rearranged to determine the length of any side of the triangle that is not known.

When the hypotenuse c is not known, use: $c^2 = a^2 + b^2$

When side a is not known, use: $a^2 = c^2 - b^2$

When side b is not known, use: $b^2 = c^2 - a^2$

Once the square of the unknown side is determined, its square root must be calculated to find the value. A **square root** is a number that produces a specified quantity when multiplied by itself. To determine a number's square root, a calculator is often needed to perform the operation by use of the square root key √. When calculating square roots, recognizing natural numbers when squared will eliminate the need for using a calculator.

Example 21-3

One side of a right triangle measures 6′ and the other side measures 8′. In this case, the hypotenuse is not known, so the formula $c^2 = a^2 + b^2$ is used.

$$c^2 = a^2 + b^2$$
$$c^2 = 6^2 + 8^2$$
$$c^2 = 36 + 64$$
$$c^2 = 100$$

Since we know that 100 is a perfect square, $c^2 = 10'$.

As squared numbers increase in size and are often not perfect squares, it becomes necessary to use a calculator and the square root key √.

Example 21-4

Side b of a right triangle measures 24′ and the hypotenuse measures 30′. In this case, side a is not known, so the formula $a^2 = c^2 - b^2$ is used.

$$a^2 = c^2 - b^2$$
$$a^2 = 30^2 - 24^2$$
$$a^2 = 900 - 576$$
$$a^2 = 324$$

Now determine the square root of the sum to find the value of a.

$$a = \sqrt{a^2}$$
$$a = \sqrt{324}$$
$$a = 18'$$

Example 21-5

Side a of a right triangle measures 12′ and the hypotenuse measures 20′. In this case, side b is not known, so the formula $b^2 = c^2 - a^2$ is used.

$$b^2 = c^2 - a^2$$
$$b^2 = 20^2 - 12^2$$
$$b^2 = 400 - 144$$
$$b^2 = 256$$

Now determine the square root of the sum to find the value of b.

$$b = \sqrt{b^2}$$
$$b = \sqrt{256}$$
$$b = 16'$$

When using the Pythagorean theorem to determine the hypotenuse of a right triangle, often the length of the hypotenuse will contain a decimal (23.74593′). The whole number represents feet and the numbers to the right of the decimal represent decimal equivalents in feet. In the construction trades, this type of an answer cannot be used in its present form since .74593′ cannot be located on a standard tape measure. When this occurs, the decimal equivalents in feet need to be converted to feet and inches. This process was covered in Unit 15, *Linear Measurement*.

Carpentry Notes

Estimating Stair Carriage Length

During the construction of a home, a deck, or any structure that requires stairs, the carpenter must lay out, cut, and install the stair carriages that will support the treads and risers. To determine what length of stock would be suitable for the carriages shown below, the Pythagorean theorem is used.

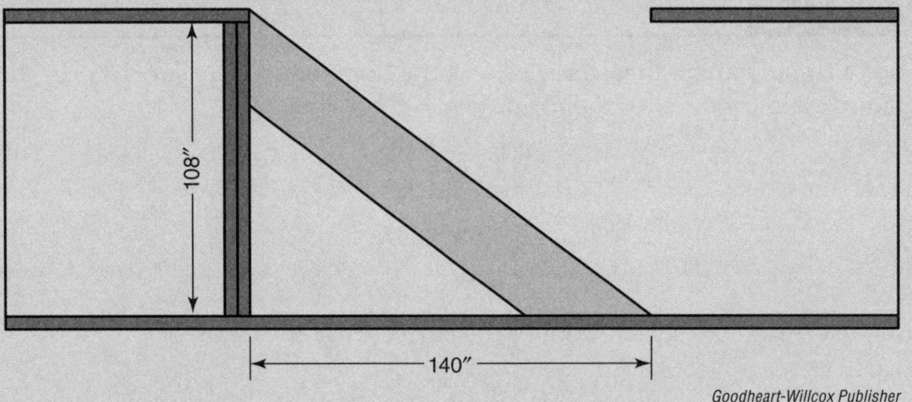

Goodheart-Willcox Publisher

The first step is to convert the inches to decimal equivalents of a foot.

Total rise 108″ = 9′

Total run 140″ = 11′ 8″ = 11.666′

Viewing the staircase as a right triangle, the carriage becomes the hypotenuse. The total rise and total run are inserted into the Pythagorean theorem as the sides.

$$c^2 = a^2 + b^2$$

$$c^2 = 9^2 + 11.666^2$$

$$c^2 = 81 + 136.095556$$

$$c^2 = 217.095556$$

Now determine the square root of the sum to find the value of c.

$$c = \sqrt{c^2}$$

$$c = \sqrt{217.095556}$$

$$c = 14.734'$$

The hypotenuse of 14.734′ does not need to be converted from decimal equivalents in feet to fractions because lumber is purchased in even lengths (12′, 14′, 16′, etc.). Accounting for angles and waste, 16′ material will be required to cut and install the stair carriages.

Unit 21 Review

Name ______________________________ Date ____________ Class ____________

Determine the length of the missing sides of a right triangle in the table below. Convert all answers to fractional feet and inches.

	Side *a*	Side *b*	Hypotenuse *c*
1.	12′	18′	
2.		15′	21′
3.	10′		18′
4.	18′	26′	
5.		22′	32′
6.	13′		23′
7.	19′ 6″	27′ 3″	
8.		14′ 9″	18′ 6″
9.	15′ 3″		27′

Goodheart-Willcox Publisher

Solve the following problems. Show all of your work.

10. A shed deck measures 12′ wide and 16′ long. What is the diagonal measurement of the floor?

____

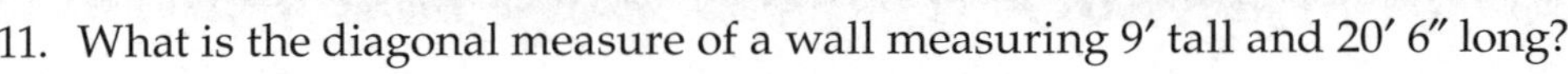

11. What is the diagonal measure of a wall measuring 9′ tall and 20′ 6″ long?

____

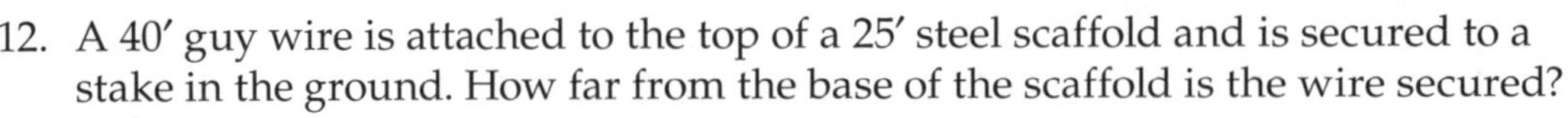

12. A 40′ guy wire is attached to the top of a 25′ steel scaffold and is secured to a stake in the ground. How far from the base of the scaffold is the wire secured?

____

13. Determine the height of the gable peak shown below.

____

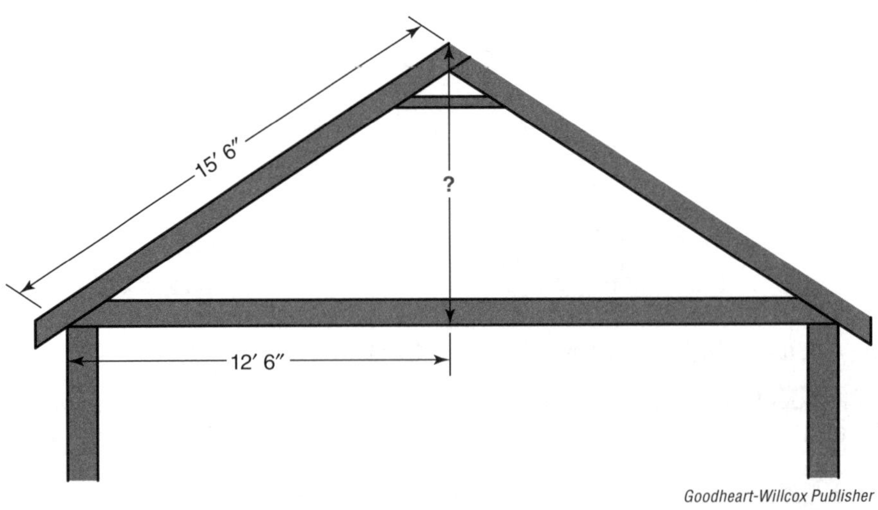

Goodheart-Willcox Publisher

Name ______________________ Date ____________ Class ____________

14. What length of straight ladder would be needed to reach the top of a 16′ 3″ high platform if the ladder is 4′ 6″ away from the platform at its base?

____

15. What is the length of the missing side of a right triangle if one side measures 36′ and the hypotenuse measures 60′?

____

16. Edge forms are being set for a 10′ × 10′ concrete patio. To ensure the forms are set square, what is the length of the diagonal?

____

17. Based on the staircase shown below, what would the length of the stair carriage material be? Round up to the nearest even foot.

____

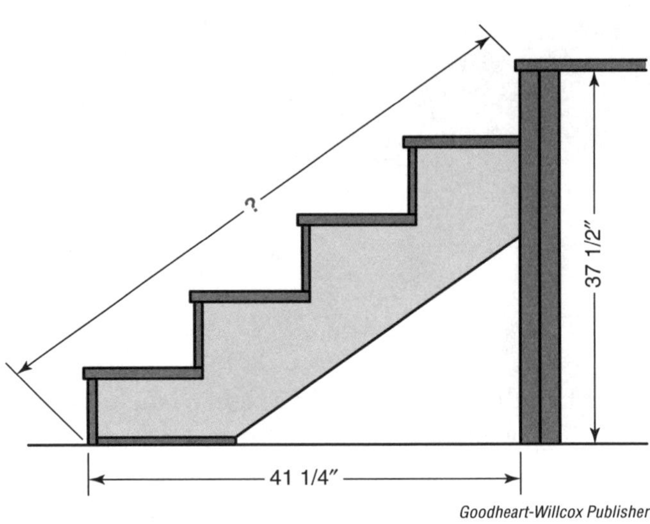

Goodheart-Willcox Publisher

18. What is the diagonal measurement of a cabinet measuring 39″ wide and 69″ tall?

____

19. What is the height of a right triangle if one side measures 100′ and the hypotenuse measures 125′?

____

Name ______________________ Date ______________ Class ______________

20. A contractor staking out the building shown below installed a stake at A, and then measured 68′ and installed a stake at B. The length of the other wall is 42′. To locate the position of the third stake at C, what is the length of the hypotenuse?

____

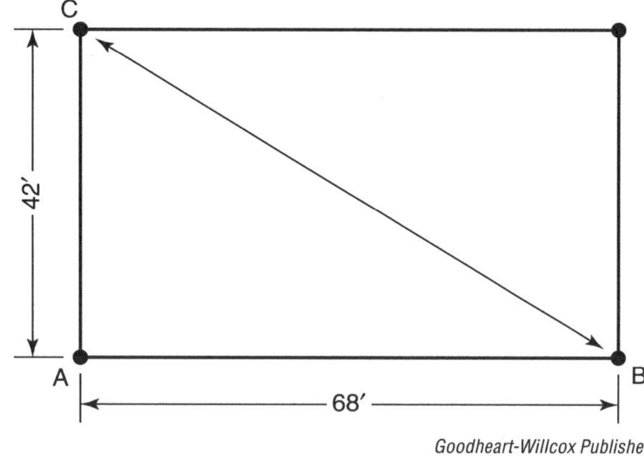

Goodheart-Willcox Publisher

Work Space/Notes

Section 5 Exam

Name ______________________ Date ____________ Class ____________

Solve the following problems showing all of your work. Convert answers to feet and fractional inches when applicable.

1. Express 55% as a fraction.

   ____

2. Express 28% as a fraction.

   ____

3. Express 85% as a decimal.

   ____

4. Express 125% as a decimal.

   ____

5. Express 0.98 as a percent.

____

6. Express 2.25 as a percent.

____

7. Express 5/8 as a percent.

____

8. Express 7/40 as a percent.

____

9. What is 16% of 180?

____

Name ______________________ **Date** __________ **Class** __________

10. 15 is what percent of 120?

____

11. 9 is 30% of what number?

____

12. Side a = 12″. Side b = 18″. What is the length of the hypotenuse?

____

13. Side a = 6′ 3″. Hypotenuse = 10′ 5″. What is the length of side b?

____

14. Side b = 15′. Hypotenuse = 18′ 9″. What is the length of side a?

____

15. Side a = 16′ 3″. Side b = 22′ 6 1/2″. What is the length of the hypotenuse?

____

16. Side a = 8′ 9″. Hypotenuse = 14′ 7″. What is the length of side b?

____

17. Side b = 14′. Hypotenuse = 17′ 6″. What is the length of side a?

____

18. A 17,000 sq ft warehouse had a 25′ × 75′ section partitioned off for office space. What percentage of the warehouse does the office space occupy?

____

Name ______________________ Date ____________ Class ____________

19. A concrete patio measures 14′ wide and 16′ 9″ long. What is the diagonal measurement of the patio?

____

20. What is the diagonal measure of a concrete form measuring 16′ wide by 18′ 3″ long?

____

21. The base of an extension ladder is positioned 6′ away from a wall that is 24′ high. What is the distance the ladder extends from the ground to the top of the wall?

____

22. Corner stakes are being set for a 38′ × 64′ excavation. To ensure the stakes are set square, what is the length of the diagonal?

____

23. The total cost of construction for a new home was $225,000. If subcontractors were contracted to perform 35% of the work, how much of the total cost was estimated for their work?

____

24. What is the diagonal measurement of a 9′ × 24′ wall?

____

25. Edge forms need to be squared while forming a 10′ 3″ × 12′ 6″ slab. What would the diagonal measure?

____

SECTION 6

Material Estimating Activities

As you become more knowledgeable and skilled in the construction trades, one skill that is essential to learn is the ability to estimate construction materials. No matter the size and scope, every project a carpenter builds utilizes materials that are purchased specifically for that project. This unit will teach you the formulas and steps used to estimate construction materials.

As your career unfolds, you will need to calculate accurate material quantities for two reasons. Initially, potential customers will receive estimates from multiple contractors to determine which one they will hire to build their project. Your ability to estimate the materials for the job and give them a competitive bid goes a long way in determining if you will be awarded the job. Secondly, when you are working on a jobsite and need to order materials for the next phase of the project, accurate calculations are critical to ensure the correct materials are delivered.

For a list of components and terms discussed in this unit, refer to Appendix A, *Construction Diagrams and Terms*.

ACTIVITY 1

Estimating Concrete for Slabs, Footers, and Walls

Objective

After studying this section, you will be able to:

- Estimate concrete for slabs, footers, and walls.

Concrete is ordered for slabs (sidewalks, patios, driveways, cellar floors), footers below a foundation wall, or vertical walls. Concrete is calculated by its volume taking into account the projects thickness (t), width (w), and length (l) in feet (see Unit 18, *Volume Measurement*). The formula used to calculate concrete quantities is:

$$\frac{t' \times w' \times l'}{27} = \text{cubic yards of concrete}$$

When the thickness, width, and length are calculated in feet, the answer will be in cubic feet. Since concrete is ordered and sold by the cubic yard, it is necessary to convert the answer to cubic yards. There are 27 cubic feet within a cubic yard.

Example 22-1

Calculate the amount of concrete needed for a project measuring 6″ thick × 14′ wide × 22′ long.

$$\frac{.5' \times 14' \times 22'}{27} = \frac{154 \text{ cu ft}}{27} = 5.7 \text{ cubic yards of concrete}$$

Math Tip

When the measurements for concrete are inserted into the formula, any variable that is not an even foot must be expressed as a decimal foot (see Unit 15, *Linear Measurement*). For example, 4″ = .333′.

Slabs

Once the measurements of a slab are determined, they are inserted into the formula and the calculations are made to determine the cubic yards needed. A good practice to follow is to add 5% onto your total to allow for possible spillage or over-excavation of the site.

Example 22-2

Calculate how much concrete is needed for a sidewalk measuring 4″ thick, 3′ wide, and 48′ long.

$$\frac{.333' \times 3' \times 48'}{27} = \frac{47.952 \text{ cu ft}}{27} = 1.776$$

$$1.776 + 5\% = 1.86$$

1.86 cubic yards of concrete is needed.

Footers

To calculate concrete for a footer, the thickness and width variables are determined by the size of the foundation wall. The footer thickness is equal to the foundation width, and the footer width is two times the foundation width. As discussed in Unit 16, *Perimeter Measurement*, the length variable for the formula is the total perimeter of the foundation.

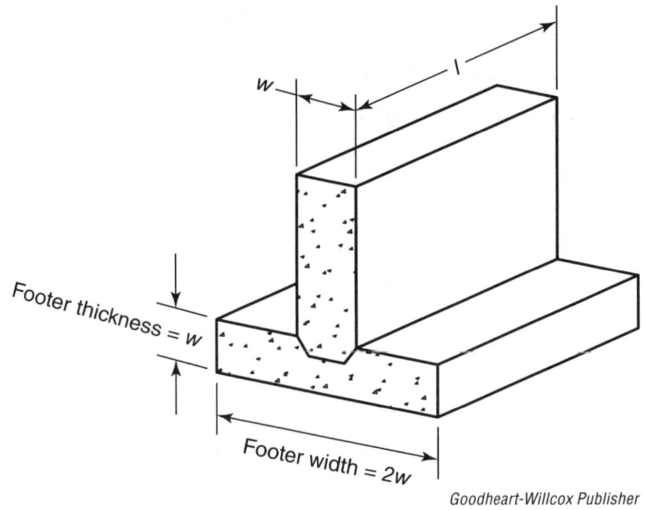

Goodheart-Willcox Publisher

Example 22-3

Calculate how much concrete should be ordered for a footer measuring 8″ thick, 16″ wide, and 112′ long.

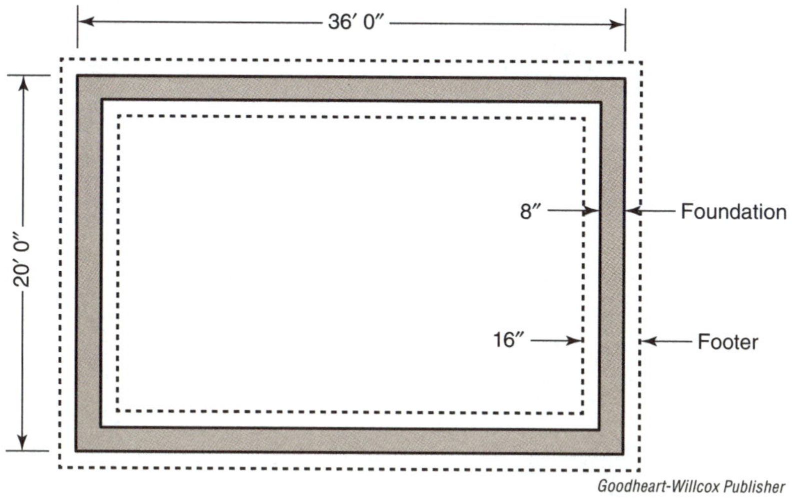

Goodheart-Willcox Publisher

$$\frac{.666' \times 1.333' \times 112'}{27} = \frac{99.431 \text{ cu ft}}{27} = 3.682$$

$$3.682 + 5\% = 3.866$$

3.866 cubic yards of concrete should be ordered.

Concrete Walls

Concrete walls are nothing more than a horizontal slab sitting vertically on edge. The thickness variable is the thickness of the wall. The width variable is the height of the wall, and the length variable is the length of the wall.

Example 22-4

Calculate how much concrete should be ordered for a wall measuring 10″ thick, 9′ 6″ high, and 28′ long.

$$\frac{.833' \times 9.5'' \times 28'}{27} = \frac{221.578 \text{ cu ft}}{27} = 8.21$$

$$8.21 + 5\% = 8.62$$

8.62 cubic yards of concrete should be ordered.

Work Space/Notes

Name ______________________ Date ____________ Class ____________

ACTIVITY 1

Estimating Concrete for Slabs, Footers, and Walls

Solve the following problems related to estimating concrete. Show all of your work and round answers to the nearest tenth.

1. Estimate concrete for a patio measuring 4″ thick × 12′ wide × 18′ long.

   ____

2. Estimate concrete for a driveway measuring 6″ thick × 22′ wide × 23′ 6″ long.

   ____

3. Estimate concrete for a sidewalk measuring 4″ thick × 3′ wide × 44′ long.

   ____

4. Estimate concrete for a basement measuring 4″ thick × 26′ 4″ wide × 46′ 4″ long.

____

5. Estimate concrete for an L-shaped sidewalk measuring 4″ thick × 3′ wide, and is 36′ long on one side and an additional 64′ long on the other side.

____

6. Estimate concrete for a footer measuring 28′ × 60′ for an 8″ foundation wall.

____

7. Estimate concrete for a footer measuring 32′ × 54′ for a 10″ foundation wall.

____

Name ______________________ Date ____________ Class ____________

8. Estimate concrete for a footer measuring 24′ × 36′. Three sides will have an 8″ foundation wall, one 36′ side will have a 12″ foundation wall.

____

9. Estimate concrete for 8′ high foundation walls on a footer that measures 28′ × 60′ × 8″ thick.

____

10. Estimate concrete for 9′ 4″ high foundation walls on a footer that measures 32′ × 54′ × 10″ thick.

____

Work Space/Notes

ACTIVITY 2

Estimating Floor Frame Materials

Objective

After studying this section, you will be able to:

- Estimate floor frame materials.

Floor systems can vary depending on architectural designs, the types of materials required, and building codes. In spite of these variations, using formulas will produce accurate results when estimating floor frame materials.

The following table shows the materials to estimate, the unit they are ordered in, and the formula used to determine the quantity. Whenever the letters "LS" appear in a formula, it refers to the *length of stock* that will be used.

Estimating Floor Frame Materials

Material	Units Ordered By	Formula
Sill seal	Lineal foot	$l + w \times 2$
Termite shield	Lineal foot	$l + w \times 2$
Sill plate	Number of boards	$l + w \times 2 \div \text{LS}$
Beams	Number of boards	$l \times \#\text{ pieces} \div \text{LS}$
Joists	Number of boards	$l \times .75 + 1 \times 2$
Header joist (boxing)	Number of boards	$l \times 2 \div \text{LS}$
Cross bridging	Number of pieces	$l \times .75 \times 4$
Subfloor	Number of sheets	$l \times w \div 32$

Goodheart-Willcox Publisher

Math Tip

Standard framing practices for floors, walls, and roofs generally place framing members at 16″ centers and occasionally at 24″ centers. When estimating on-center components within these systems, .75 is the variable used for 16″ centers and .5 is the variable used for 24″ centers. When determining the number of joists, studs, or rafters, the formula always contain a +1 variable. This accounts for the component on the end where your measuring tape reads zero.

Example 22-5

Estimate materials needed for a 28′ × 60′ floor system.

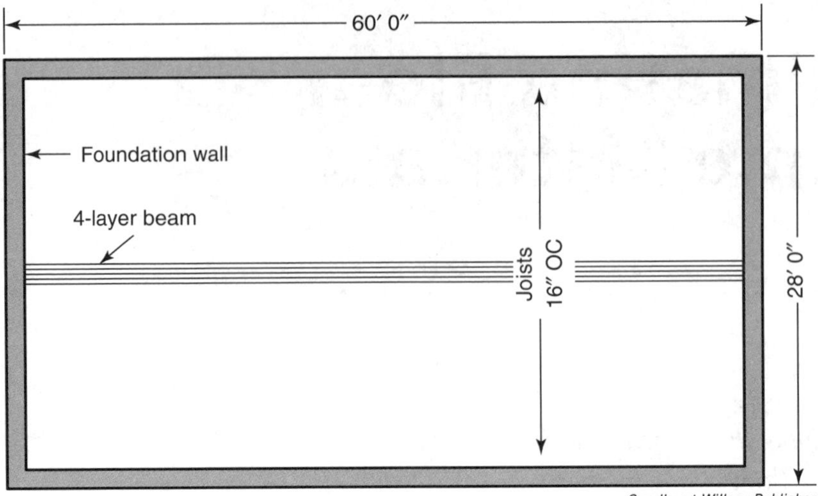

Goodheart-Willcox Publisher

Part	Formula	Material Estimate
Sill seal	$l + w \times 2$	60 + 28 × 2 = **176 lineal feet**
Termite shield	$l + w \times 2$	60 + 28 × 2 = **176 lineal feet**
Sill plate (2 × 6 × 14′)	$l + w \times 2 \div$ LS	60 + 28 × 2 ÷ 14 = 12.57 = **13 boards**
Beams (2 × 12 × 16′)	$l \times$ # pieces ÷ LS	60 × 4 ÷ 16 = **15 boards**
Joists (2 × 10 × 14′)	$l \times .75 + 1 \times 2$	60 × .75 + 1 × 2 = **92 boards**
Header joist (2 × 10 × 14′) *Note: Header joists for the 28′ sides were factored in with the joists.*	$l \times 2 \div$ LS	60 × 2 ÷ 14 = 8.57 = **9 boards**
Cross bridging *Note: The × 4 accounts for 2 pieces per cavity on both sides of the beam.*	$l \times .75 \times 4$	60 × .75 × 4 = **180 pieces**
Subfloor	$l \times w \div 32$	60 × 28 ÷ 32 = 52.5 = **53 sheets**

Goodheart-Willcox Publisher

Name ________________ Date ________ Class ________

ACTIVITY 2

Estimating Floor Frame Materials

Use the following floor plan to calculate questions 1–8 related to floor plan estimations. Show all your work.

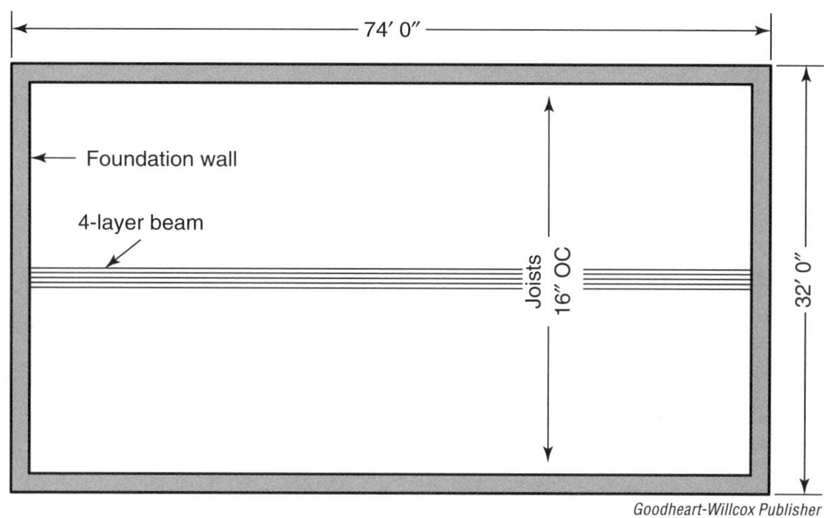

Goodheart-Willcox Publisher

1. Sill seal ________ lineal feet

2. Termite shield ________ lineal feet

3. Sill plate ($2 \times 6 \times 16'$) ________ boards

4. Beams ($2 \times 10 \times 14'$) ________ boards

5. Joists ($2 \times 10 \times 16'$) ________ boards

6. Header joist ($2 \times 10 \times 16'$) ________ boards

Name ________________________ **Date** ____________ **Class** ____________

7. Cross-bridging _______ pieces

8. Subfloor _______ sheets

Estimate the materials needed to construct a floor system on a foundation measuring 26′ wide × 48′ long framed at 16″ centers. It will have a built-up 2 × 12 beam 3 members wide. Use this information to solve questions 9–16. Show all of your work.

9. Sill seal _______ lineal feet

10. Termite shield _______ lineal feet

11. Sill plate (2 × 6 × 16′) ________ boards

12. Beams (2 × 10 × 14′) ________ boards

13. Joists (2 × 10 × 14′) ________ boards

14. Header joist (2 × 10 × 16′) ________ boards

15. Cross-bridging ________ pieces

Name ______________________________ **Date** ______________ **Class** ______________

16. Subfloor _______ sheets

Estimate the materials needed to construct a floor system on a foundation measuring 30′ wide × 56′ long framed at 24″ centers. It will have a built-up 2 × 10 beam 4 members wide. Use this information to solve questions 17–24. Show all of your work.

17. Sill seal _______ lineal feet

18. Termite shield _______ lineal feet

19. Sill plate (2 × 6 × 16′) _______ boards

20. Beams ($2 \times 10 \times 14'$) ________boards

21. Joists ($2 \times 10 \times 16'$) ________boards

22. Header joist ($2 \times 10 \times 16'$) ________boards

23. Cross-bridging ________pieces

24. Subfloor ________sheets

ACTIVITY 3

Estimating Wall Frame Materials

Objective

After studying this section, you will be able to:

- Estimate wall frame materials.

Wall frame estimates are calculated for every wall within a structure because of their individual unique qualities. These include their size, location (exterior wall or interior partition), components within the wall (window/door rough openings, corner post, partition post), and on-center spacing. All these factors are taken into consideration as you use the wall estimation formulas.

The following table shows the materials needed when estimating a wall frame, the units they are ordered in, and the formula used to determine the quantity.

Estimating Wall Frame Materials

Material	Units Ordered	Formula
Sole and double top plates	# of boards	$l \times 3 \div$ LS
Studs	# of boards	$l \times .75 + 1 + 2$ (RO, PP) + 1 CP
Exterior sheathing	# of sheets	$l \times w \div 32$

Goodheart-Willcox Publisher

Example 22-6

Before calculating a wall, first assess the components within the wall and the on-center spacing. The wall shown here contains 4 rough openings (RO), 2 partition posts (PP), 2 corner posts (CP), and is to be built 16″ on-center (OC).

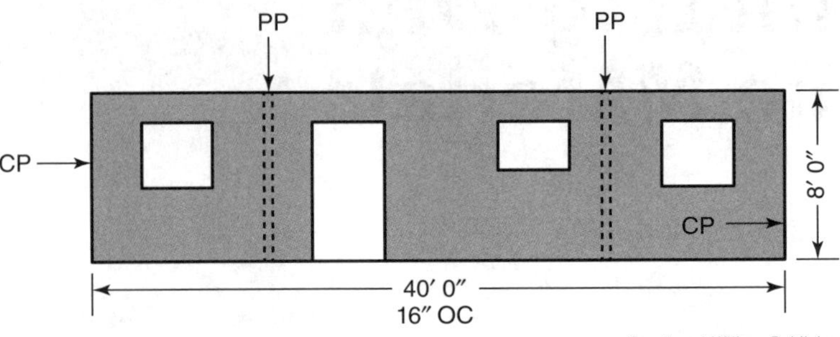

Goodheart-Willcox Publisher

Part	Formula	Material Estimate
Sole and double top plates (2 × 4 × 16′)	*l* × 3 ÷ LS	40 × 3 ÷ 16 120 ÷ 16 = 7.5 = **8 boards**
Studs (2 × 4 × 8′)	*l* × .75 + 1 + 2 (RO, PP) + 1 CP	40 × .75 + 1 + 2 (4 RO, 2 PP) + 2 CP 30 + 1 + 2(6) + 2 31 + 12 + 2 = **45 boards**
Exterior sheathing (7/16″ OSB)	*l* × w ÷ 32	40 × 8 ÷ 32 320 ÷ 32 = **10 sheets**

Goodheart-Willcox Publisher

The header material is the only item not included in this estimate. The size of header material is based on the span of the rough opening; therefore, all headers cannot be calculated together. Use the following steps to estimate header material for a rough opening:

1. Determine the size of stock based on rough opening span charts.

   2 × 8 stock

2. Determine the rough opening width and add 3″ for trimmer studs, then round up to the nearest foot.

   32″ RO + 3″ = 35″ = 3′

3. Double this number for a 2 × 4 wall; triple it for a 2 × 6 wall.

   3′ × 2 = 6′   2 × 8 × 6′ for a 2 × 4 wall

   3′ × 3 = 9′   2 × 8 × 9′ for a 2 × 6 wall

Name ______________________ Date __________ Class __________

ACTIVITY 3

Estimating Wall Frame Materials

Use the following image to answer questions 1–3 regarding estimating materials for a wall frame. Show all of your work.

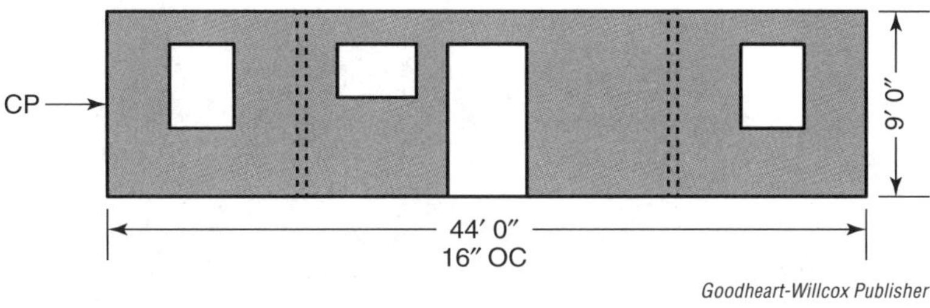

Goodheart-Willcox Publisher

1. Sole and double top plates (2 × 4 × 16′) _______ boards

2. Studs (2 × 4 × 8′) _______ boards

3. Exterior sheathing (7/16″ OSB) _______ sheets

Calculate the materials to construct a 40′ long × 8′ high wall framed 16″ on center. A corner post is required at both ends of the wall and it will have three intersecting partitions. Rough openings consist of a 3′ × 7′ door, a 2′ × 3′ window, and two 3′ × 3′ windows. Use this information to solve questions 4–6. Show all of your work.

4. Sole and double top plates (2 × 4 × 16′) ________ boards

5. Studs (2 × 4 × 8′) ________ boards

6. Exterior sheathing (7/16″ OSB) ________ sheets

Name ______________________________ **Date** ____________ **Class** ____________

Calculate the materials to construct a 54′ long × 8′ high wall framed 24″ on center. A corner post is required on one end of the wall and it will have two intersecting partitions. Rough openings consist of a 6′ × 7′ door and four 3′ × 4′ windows. Use this information to solve questions 7–9. Show all of your work.

7. Sole and double top plates (2 × 6 × 16′) _______ boards

8. Studs (2 × 6 × 8′) _______ boards

9. Exterior sheathing (7/16″ OSB) _______ sheets

10. Use the following image to estimate how many sheets of 7/16″ OSB are needed to sheath the end wall.

____

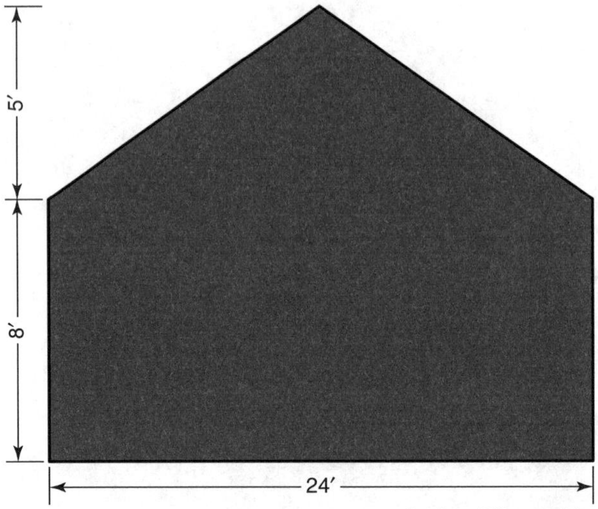

Goodheart-Willcox Publisher

ACTIVITY 4

Estimating Ceiling Frame Materials

Objective

After studying this section, you will be able to:

- Estimate ceiling frame materials.

A first floor ceiling on a two-story home also serves as the floor system for the second floor. In this case, floor estimation practices are used. This section discusses the second floor ceiling of a two story or the ceiling of a single story home. With the exception of ceiling joists and subflooring, other components utilized in floor framing are not used in a ceiling system. Ceiling joists are calculated just like floor joists using the same formula. On occasions, subflooring is estimated and installed on a section of a ceiling for attic storage or future living space. The same formula used in subfloors is then applied. For example, the number of ceiling joists for a 28′ × 60′ floor system would be 92 boards.

$$2 \times 8 \times 14', 16'' \text{ OC}$$

$$1 \times .75 + 1 \times 2$$

$$60 \times .75 + 1 \times 2 = \textbf{92 boards}$$

Work Space/Notes

Name ______________________ Date ____________ Class ____________

ACTIVITY 4

Estimating Ceiling Frame Materials

Estimate the materials needed to construct a ceiling system on a building measuring 22′ wide × 44′ long framed at 16″ centers. The subfloor will cover a 16′ × 44′ area. Show all of your work.

1. Joists (2 × 8 × 12′) ________ boards

2. Subfloor ________ sheets

Estimate the materials needed to construct a ceiling system on a building measuring 26′ × 52′ long framed at 24″ centers. The subfloor will cover a 12′ × 52′ area. Show all of your work.

3. Joists (2 × 8 × 14′) ________ boards

4. Subfloor ________ sheets

Estimate the materials needed to construct a ceiling system on a building measuring 25′ wide × 42′ long framed at 16″ centers. The subfloor will cover a 24′ × 36′ area. Show all of your work.

5. Joists (2 × 10 × 14′) ________ boards

6. Subfloor ________ sheets

Estimate the materials needed to construct a ceiling system on a building measuring 30′ wide × 64′ long framed at 24″ centers. The subfloor will cover a 26′ × 48′ area. Show all of your work.

7. Joists (2 × 10 × 16′) ________ boards

8. Subfloor ________ sheets

Name ______________________________ **Date** ______________ **Class** ______________

Estimate the materials needed to construct a ceiling system on a building measuring 22′ wide × 40′ long framed at 16″ centers. The subfloor will cover a 16′ × 40′ area. Show all of your work.

9. Joists (2 × 8 × 12′) ________ boards

10. Subfloor ________ sheets

Work Space/Notes

ACTIVITY 5

Estimating Roof Frame Materials

Objective

After studying this section, you will be able to:

- Estimate roof frame materials.

Of all the framing systems within a structure, roof systems are the most complex. Multiple factors must be considered, such as the style of the roof, the pitch, and the size of the overhangs to name just a few.

The most common type of roof used in construction is the gable roof. The following steps and formulas will explain how to calculate materials for a gable roof. Adaptations to the formulas can be made to account for other roof styles and designs.

Before the estimation process begins, you must first calculate the overall line length of the rafter including the tail. This will determine the length of the roof on the slope and the length of the rafter material.

Estimating Gable Roof Frames

Material	Units Ordered	Formula
Rafters	# of boards	BL × .75 + 3 × 2
Ridge board	# of boards	RL ÷ LS
Roof sheathing	# of sheets	RL × RS × 2 ÷ 32

BL = building length; RS = roof slope; RL = roof length; LS = length of stock

Goodheart-Willcox Publisher

Example 22-7

Estimate materials for a 20′ × 32′ gable roof.

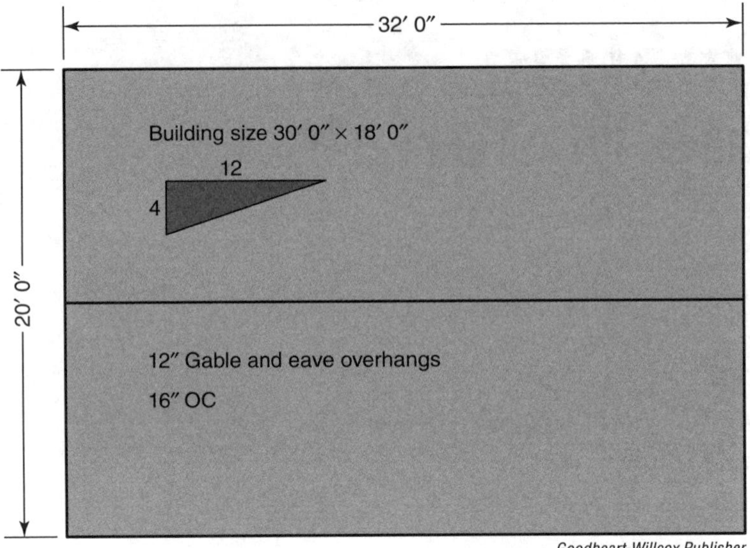

Goodheart-Willcox Publisher

To determine the common rafter length, use the following formula:

Unit length from rafter table × total run (including the overhang)

12.65 × 10 = 126.5 or 10′ 6 1/2″

5	4	3	2
13.00	12.65	12.37	12.16
17.69	17.44	17.23	17.09
17.33	16.87	16.49	16.22
26	25.30	24.74	24.33
11 1/16	11 3/8	11 5/8	11 13/16
11 1/2	11 11/16	11 13/16	11 15/16

Goodheart-Willcox Publisher

Round the answer up to the nearest even foot (12′ 0″). This will be the length of the rafter material.

Part	Formula	Material Estimate
Rafters (2 × 6 × 12′)	BL × .75 + 3 × 2	30 × .75 + 3 × 2 30 × .75 = 22.5 22.5 + 3 × 2 = 51 **52 boards** *Note: This must always be an even number.*
Ridge board (2 × 8 × 16′)	RL ÷ LS	32 ÷ 16 = 2 boards
Roof sheathing	RL × RS × 2 ÷ 32	32 × 12 × 2 ÷ 32 768 ÷ 32 = **24 sheets**

Goodheart-Willcox Publisher

Name ______________________ Date ____________ Class ____________

ACTIVITY 5

Estimating Roof Frame Materials

Create a roof frame estimate. Use the following images to solve questions 1–4.

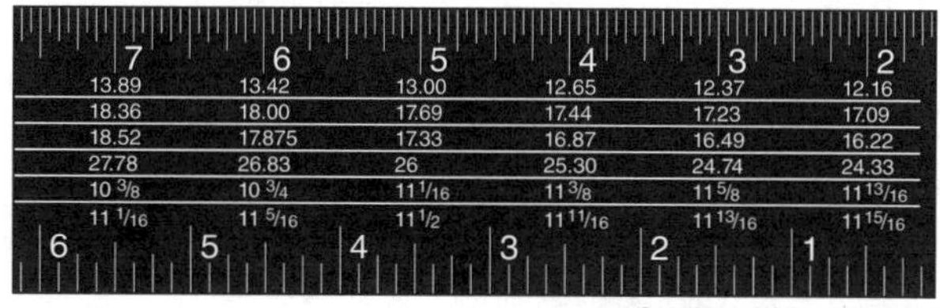

Goodheart-Willcox Publisher

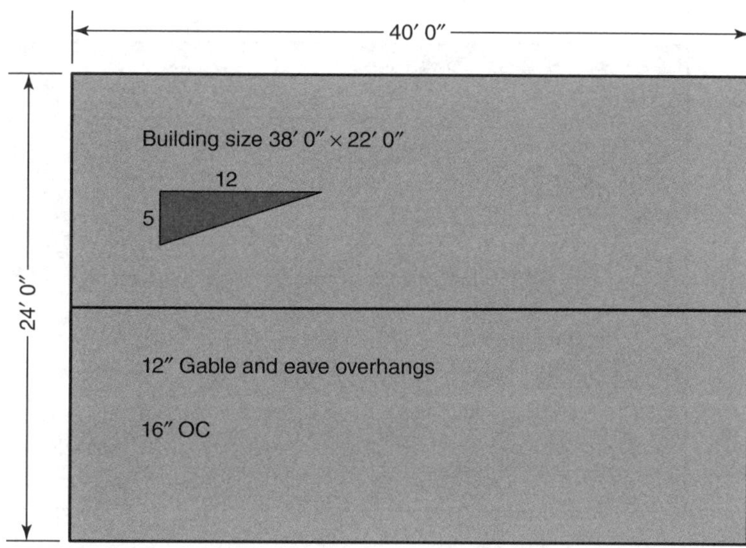

Goodheart-Willcox Publisher

1. Line length of the rafter including the tail ________

2. Rafters (2 × 6 × ________′) ________ boards

3. Ridge board (2 × 8 × 16′) ________ boards

4. Roof sheathing ________ sheets

Create a roof frame estimate. Use the following images to solve questions 5–8.

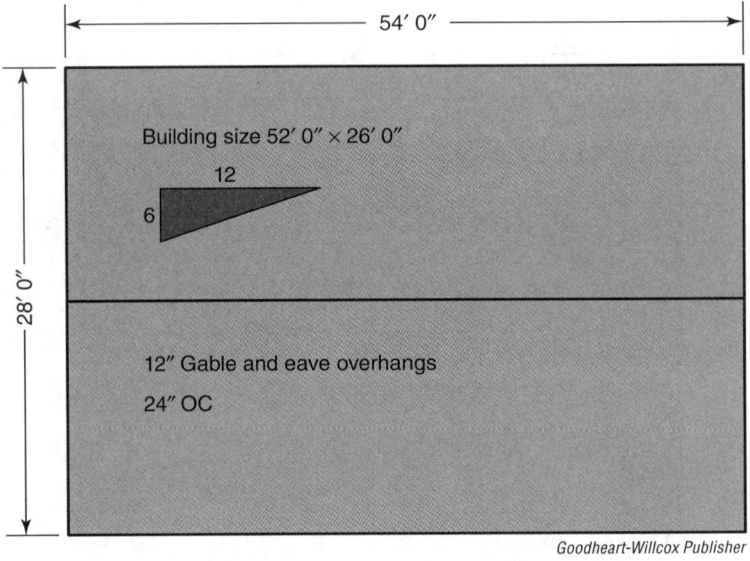

Goodheart-Willcox Publisher

5. Line length of the rafter including the tail ________

6. Rafters (2 × 8 × ________′) ________ boards

Name ______________________ **Date** __________ **Class** __________

7. Ridge board ($2 \times 10 \times 16'$) ________ boards

8. Roof sheathing ________ sheets

Create a roof frame estimate. Use the following images to solve questions 9–12.

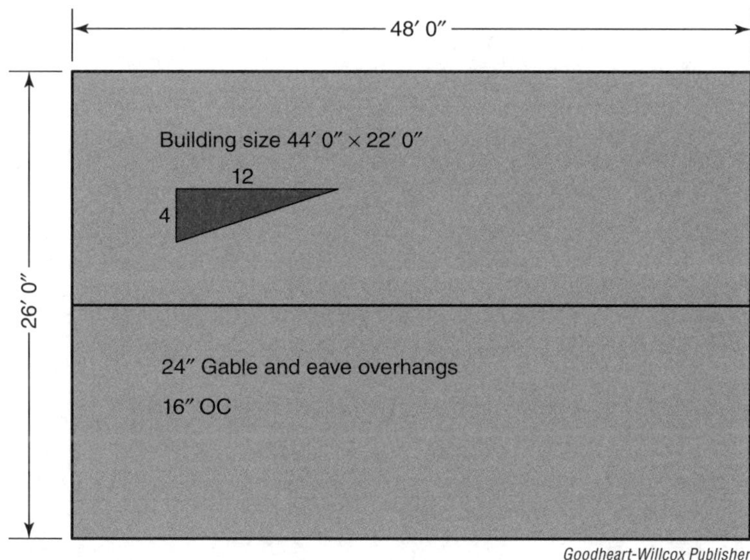

Goodheart-Willcox Publisher

9. Line length of the rafter including the tail ________

10. Rafters (2 × 6 × ________) ________ boards

11. Ridge board (2 × 8 × 16′) ________ boards

12. Roof sheathing ________ sheets

ACTIVITY 6

Estimating Roof Finish Materials

Objective

After studying this section, you will be able to:

- Estimate roof finish materials.

Calculating roof finish materials can be as complex as the roof systems they cover. With multiple types of roof styles used in construction, this instruction will cover estimation process for a gable roof system using three-tab shingles. Generally, shingle designs range from 3 to 4 bundles per square. Adaptations to the formulas can be made to account for other roof styles and designs. Additional shingles are needed for the starter course and caps. These are accounted for by adding a standard 5% waste factor. When roofs become more complex with hips/valleys and dormers, it is common to start adding an additional 10% to the waste factor.

The following table shows the roof finish estimate for material for a gable roof, the unit they are ordered in, and the formula used to determine the quantity.

Estimating Roof Finish Materials

Material	Units Ordered By	Formula
Drip edge	# pieces	l (2) + slope(4) ÷ 10
Felt paper	sq ft / roll	2 (l × slope) + 5%
Shingles	# of squares	2 (l × slope) + 5% ÷ 100

Drip edge is made in 10′ lengths.

Goodheart-Willcox Publisher

Example 22-8

Estimate materials for a 42′ long roof with a 16′ slope.

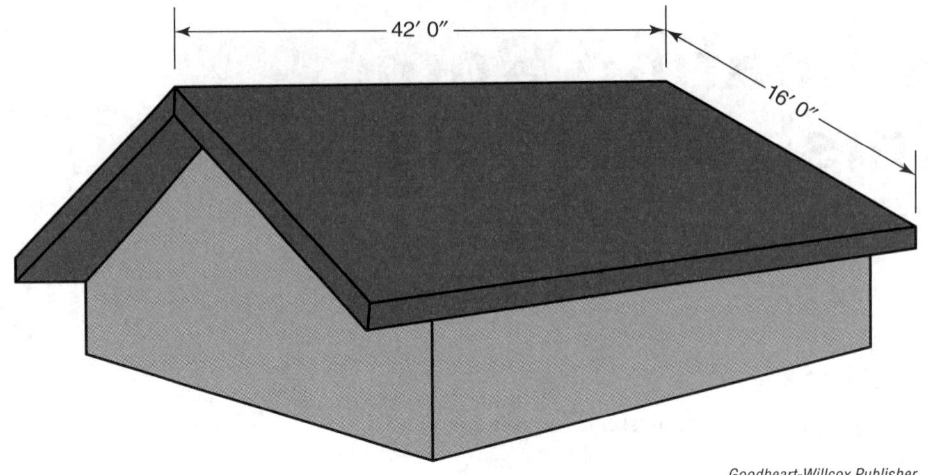

Goodheart-Willcox Publisher

Part	Formula	Material Estimate
Drip edge	*l* (2) + slope (4) ÷ 10	42 (2) + 16 (4) ÷ 10 84 + 64 ÷ 10 148 ÷ 10 = 14.8 = **15 pieces**
Felt paper	2 (*l* × slope) + 5%	2 (42 × 16) + 5% 2 × 672 + 5% 1,344 + 5% = 1,411.2 = **1,412 sq ft**
Shingles	2 (*l* × slope) + 5% ÷ 100	2 (42 × 16) + 5% ÷ 100 2 × 672 + 5% ÷ 100 1,344 + 5% = 1,411.2 1,412 ÷ 100 = **14.12 squares**

Goodheart-Willcox Publisher

Name ______________________ Date ____________ Class ____________

ACTIVITY 6

Estimating Roof Finish Materials

Use the following figure to estimate materials for a 32′ long roof with a 14′ 6″ slope. Show all of your work.

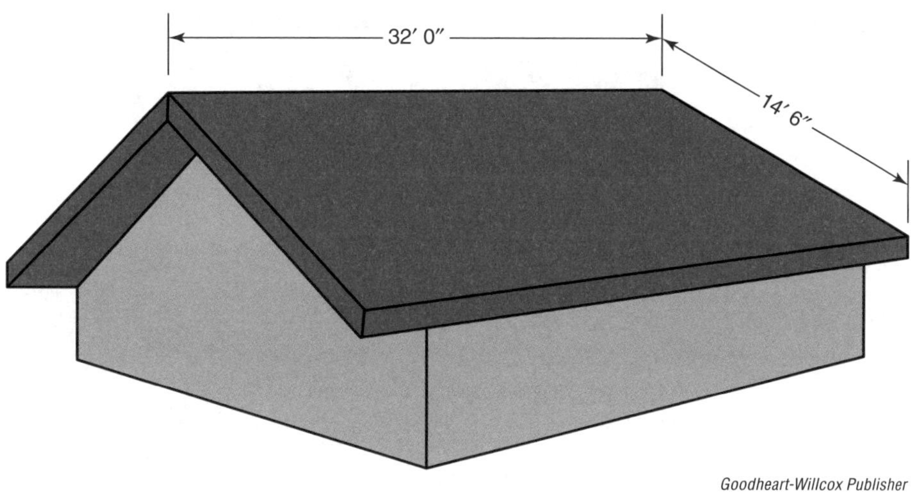

Goodheart-Willcox Publisher

1. Drip edge ________ pieces

2. Felt paper ________ square feet

3. Shingles ________ squares

Estimate the amount of roofing materials that will be needed for a roof that is 48′ long and has a 15′ slope from ridge to eave. Show all of your work.

4. Drip edge _______ pieces

5. Felt paper _______ square feet

6. Shingles _______ squares

Estimate the amount of roofing materials that will be needed for a roof that is 40′ long and has a 17′ slope from ridge to eave. Show all of your work.

7. Drip edge _______ pieces

Name ______________________________ **Date** ____________ **Class** ____________

8. Felt paper _______ square feet

9. Shingles _______ squares

Estimate the amount of roofing materials that will be needed for a roof that is 74′ long and has a 19′ slope from ridge to eave. Show all of your work.

10. Drip edge _______ pieces

11. Felt paper _______ square feet

12. Shingles _______ squares

Work Space/Notes

ACTIVITY 7

Estimating Siding Materials

Objective

After studying this section, you will be able to:

- Estimate siding materials.

There is a wide variety of siding materials and components on the market today. These include horizontal wood and vinyl, vertical wood, wood panels, concrete panels, and wood shingles and shake to name just a few.

Each type of siding has its own unique trims and accessories suited to that particular style. Horizontal vinyl siding has become the most widely used siding product because of its durability and maintenance free qualities. Refer to Appendix A, *Construction Diagrams and Terms* for a siding diagram that indicates where each component is used.

The following table shows the materials to include when estimating siding, the units they are ordered in, and the formula used to determine the quantity.

Estimating Siding Materials

Materials	Unit Ordered By	Formula
12′ 6″ F-channel	# pieces	lineal measurement
12′ Finish fascia	# pieces	lineal measurement
Horizontal siding	# squares	$l \times w \div 100$
10′ Inside corner post	# pieces	1 per floor level
12′ 6″ J channel	# pieces	lineal measurement
10′ Outside corner post	# pieces	1 per floor level
12′ Soffit	# pieces	size′ × lineal foot
10′ Starter strip	# pieces	lineal measurement
12′ 6″ Undersill trim	# pieces	lineal measurement

Goodheart-Willcox Publisher

Math Tip

When estimating siding components, it is good practice to round all measurements up to the nearest foot to account for waste.

Example 22-9

Estimate the siding and components to finish the two walls in the diagram shown here.

The door measures 3′ 6″ × 7′ 0″

The windows measure 2′ 6″ × 3′ 0″

The eave and rakes measure 12″

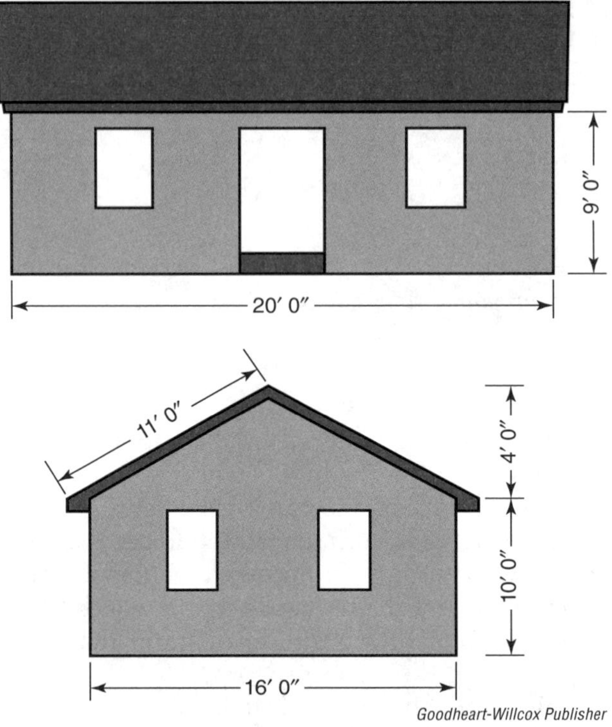

Goodheart-Willcox Publisher

Part	Material Estimate
F-channel	20′ + 11′ + 11′ = 42′ ÷ 12.5 = 3.36 = **4 pieces**
Finish fascia	22 + 11′ + 11′ = 44′ ÷ 12 = 3.66 = **4 pieces**
Horizontal siding	(20′ × 9′) + (16′ × 10′) + (4′ × 8′) 180 sq ft + 160 sq ft + 32 sq ft = 372 sq ft 372 ÷ 100 = 3.72 = **4 squares**
J channel	18′ (door) + 36′ (4 windows) + 22′ (gable) = 76′ 76 ÷ 12.5 = 6.08 = **7 pieces**
Outside corner post	**4 pieces**
Soffit	1′ (22 + 11 + 11) = 44′ ÷ 12 = 3.66 = **4 pieces**
Starter strip	20 – 3′ 6″ (door) + 16 = 32.5 ÷ 10 = 3.25 = **4 pieces**
Undersill trim	10′ (4 windows) + 20′ = 30′ ÷ 12.5 = 2.4 = **3 pieces**

Goodheart-Willcox Publisher

Name ______________________________ Date ____________ Class ____________

ACTIVITY 7

Estimating Siding Materials

Use the following information to estimate the amount of siding and components needed to finish the two walls shown. Show all of your work.

- The door measures 5′ 6″ × 7′ 0″
- Five windows measure 3′ 6″ × 4′ 0″
- One window measures 2′ 6″ × 3′ 0″

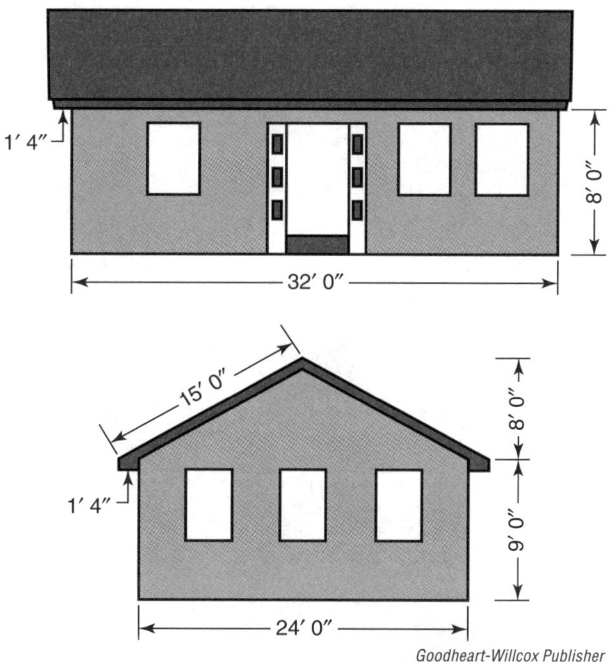

Goodheart-Willcox Publisher

1. F-channel ________ pieces

2. Finish fascia ________ pieces

3. Horizontal siding ________ squares

4. J channel ________ pieces

5. Outside corner post ________ pieces

6. Soffit ________ pieces

Name ______________________ **Date** ____________ **Class** ____________

7. Starter strip _______ pieces

8. Undersill trim _______ pieces

Use the following information to estimate the amount of siding and components needed to finish the wall shown in the illustration.

- The door measures 3′ 6″ × 7′ 0″
- Two windows measure 3′ 0″ × 4′ 0″
- One window measures 6′ 0″ × 4′ 0″
- One window measures 2′ 0″ × 2′ 0″

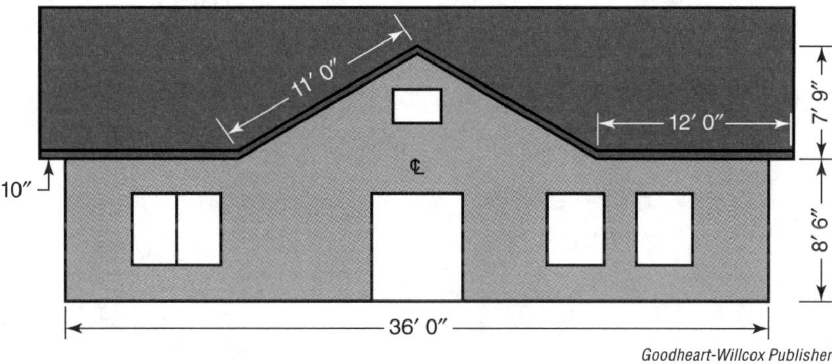

Goodheart-Willcox Publisher

9. F-channel _______ pieces

10. Finish fascia ________ pieces

11. Horizontal siding ________ squares

12. J channel ________ pieces

13. Outside corner post ________ pieces

14. Soffit ________ pieces

Name ______________________ **Date** ____________ **Class** ____________

15. Starter strip ________ pieces

16. Undersill trim ________ pieces

Work Space/Notes

Estimating Insulation Materials

Objective

After studying this section, you will be able to:

- Estimate insulation materials.

Insulation products commonly used in residential construction include blanket, batt, loose fill, and rigid. These products are manufactured in rolls, bundles, bags, and panels. Each of these units covers a specified square foot area. Panels are calculated by determining the square foot coverage per panel. State and local codes determine the minimum R-value requirements for their location. When estimating insulation, always make sure you meet these minimum requirements.

The label on the bundle of insulation shown below identifies this product as R-30 unfaced batts. When installed, this bundle will cover 88 square foot. To estimate how many bundles to purchase, first determine the square footage of the area to be insulated and then divided by 88. When estimating for walls, always include the thickness of the floor system so the cavities between the joists along the boxing will be insulated.

Goodheart-Willcox Publisher

Example 22-10

Estimate insulation for the basement, walls, and ceiling for the following 28′ × 56′ home.

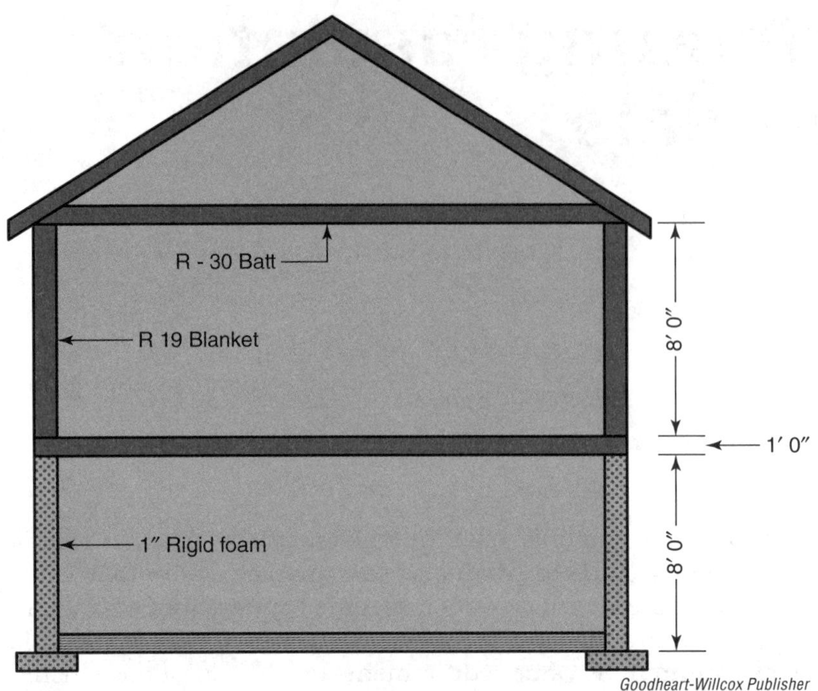

Goodheart-Willcox Publisher

Location	R Value	Square Foot per Unit	Material Estimate
Exterior walls (4)	R 19	75 sq ft/roll	28′ + 28′ + 56′ + 56′ = 168 168 × 9 = 1,512 sq ft 1,512 ÷ 75 = 20.16 = **21 rolls**
Ceiling	R 30	88 sq ft/bundle	28′ × 56′ = 1,568 sq ft 1,568 ÷ 88 = 17.8 = **18 bundles**
Cellar walls (4)	R 5	32 sq ft/panel	28′ + 28′ + 56′ + 56′ = 168 168 × 8 = 1,344 sq ft 1,344 ÷ 32 = **42 panels**

Goodheart-Willcox Publisher

Name ______________________________ Date ____________ Class ____________

ACTIVITY 8

Estimating Insulation Materials

Use the illustration shown here to estimate the amount of insulation needed for the floors, walls, and ceiling in a home that is 28′ × 52′. Show all of your work.

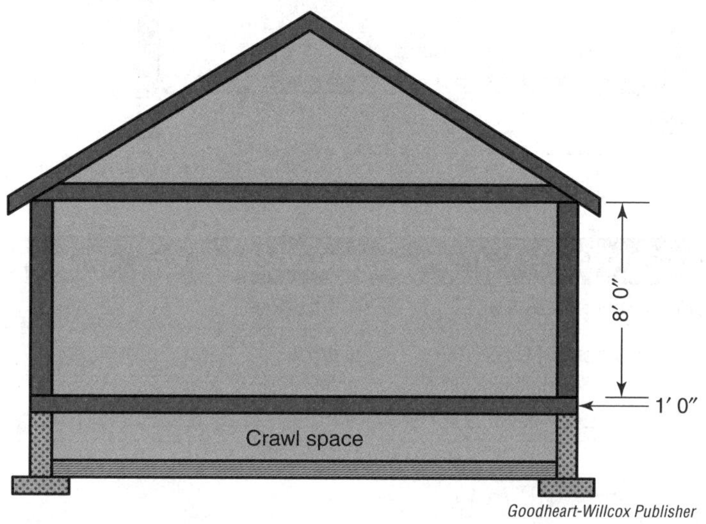

Goodheart-Willcox Publisher

	Location	R Value	Square Foot per Unit	Material Estimate
1.	Floor	R 30	88 sq ft/bundle	_______ **bundles**
2.	Four exterior walls	R 19	75 sq ft/roll	_______ **rolls**
3.	Ceiling	R 38	64 sq ft/bundle	_______ **bundles**

Goodheart-Willcox Publisher

Use the illustration to estimate the amount of insulation needed for the walls and cathedral ceiling in a 24′ × 48′ home.

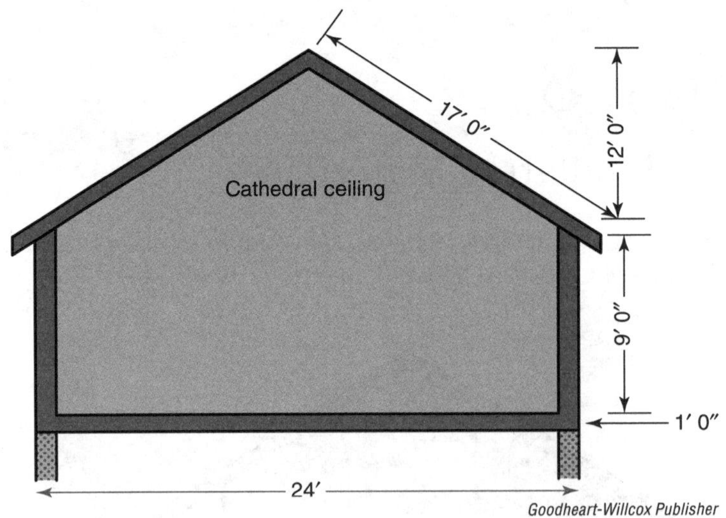

Goodheart-Willcox Publisher

	Location	R Value	Square Foot per Unit	Material Estimate
4.	Four exterior walls	R 19	75 sq ft/bundle	_______ rolls
5.	Two exterior gables	R 19	75 sq ft/roll	_______ rolls
6.	Cathedral ceiling	R 30	88 sq ft/bundle	_______ bundles

Goodheart-Willcox Publisher

Name ______________________ **Date** __________ **Class** __________

Use the illustration to estimate the amount of insulation needed for the floors, walls, and ceiling in a 26′ × 44′ home.

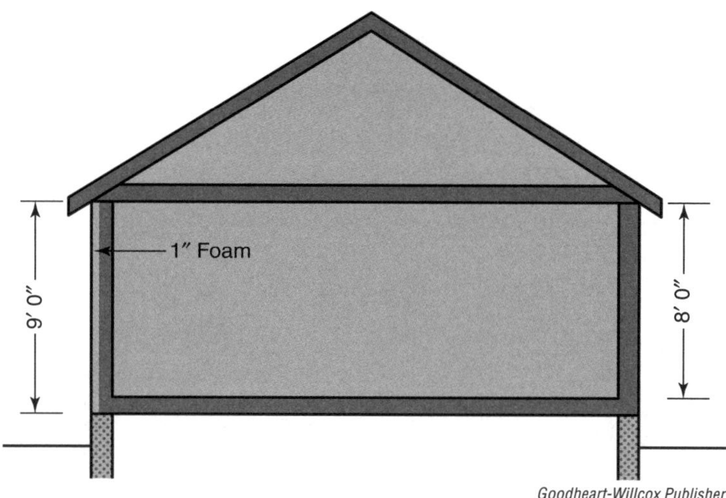

Goodheart-Willcox Publisher

	Location	R Value	Square Foot per Unit	Material Estimate
7.	Four exterior walls	R 13	107 sq ft/roll	_______ rolls
8.	Four exterior walls (including the header joist)	1″ foam	32 sq ft/panel	_______ panels
9.	Ceiling	R 38	64 sq ft/bundle	_______ bundles

Goodheart-Willcox Publisher

Work Space/Notes

ACTIVITY 9

Estimating Interior Trim Materials

Objective

After studying this section, you will be able to:

- Estimate interior trim materials.

When estimating interior trim materials, it is important to remember that running trims are purchased by the lineal foot. A specific length for window and door casings are purchased to avoid waste and splices. It is important to get into the habit of rounding up lineal foot estimates to the nearest foot for waste, and partial piece estimates to the next whole piece.

Math Tip

For standard 6′ 8″ interior doors, carpenters will generally purchase five 7′ pieces of casing to trim both sides of the door.

The following table shows the materials needed to estimate interior trim projects, the unit they are ordered in, and the formula used to determine the quantity.

Estimating Interior Trim Materials

Material	Units Ordered	Formula
Door casing (7′ length)	Number of pieces	$2\,(w + 6'') + 4\,(l + 4'')$
Window casing/apron	Number of pieces	$2\,(w + 8'') + 2\,(l + 4'')$
Window stool	Number of pieces	$w + 12$
Base molding	Lineal foot	Perimeter – door widths
Crown molding	Lineal foot	Perimeter
Chair rail	Lineal foot	Perimeter – (door + window widths)

Goodheart-Willcox Publisher

Example 22-11

Estimate the interior trim for an 18′ 6″ × 16′ 0″ room with two 3′ 0″ × 4′ 0″ windows, two 2′ 6″ × 6′ 8″ interior doors, and one 3′ 0″ × 6′ 8″ exterior door.

Location	Formula	Material Estimate
Door casing Interior doors (both sides)	2 (*w* + 6″) + 4 (*l* + 4″)	2 (2′ 6″ + 6″) + 4 (6′ 8″ + 4″) 2 (3′) + 4 (7′) 6′ + 28′ = 34′ 34′ × 2 doors = **68′ (10 – 7′ pieces**)
Door casing Exterior door (one side)	(*w* + 6″) + 2 (*l* + 4″)	(3′ 0″ + 6″) + 2 (6′ 8″ + 4″) (3′ 6″) + 2 (7′) 3′ 6″ + 14′ = **17′ 6″ (3 – 7′ pieces)**
Window casing	2 (*w* + 8″) + 2 (*l* + 4″)	2 (3′ + 8″) + 2 (4′ + 4″) 2 (3′ 8″) + 2 (4′ 4″) 7′ 4″ + 8′ 8″ (Round up to eliminate any splices in casing) 8′ and 9′ **2 – 8′ and 2 – 9′ pieces for 2 windows**
Window stool	*w* + 12″	3′ + 12″ = 4′ × 2 windows = **8′ pieces**
Base molding	Perimeter - door widths	2 (18′ 6″) + 2 (16′) – door widths 37′ + 32′ = 69′ 69′ – 8′ = **61 lineal feet**
Crown molding	Perimeter	2 (18′ 6″) + 2 (16′) = **69 lineal feet**
Chair rail	Perimeter (door + window widths)	2 (18′ 6″) + 2 (16′) 69′ – 14 = **55 lineal feet**

Goodheart-Willcox Publisher

Name ______________________ Date ____________ Class ____________

ACTIVITY 9

Estimating Interior Trim Materials

Estimate the interior trim materials needed for a 15′ 6″ × 18′ 3″ room with two 3′ 0″ × 5′ 0″ windows, one 2′ 6″ × 6′ 8″ interior door, and one 3′ 0″ × 6′ 8″ exterior door. Show all of your work.

1. Exterior door casing ________, 7′ pieces

2. Interior door casing ________, 7′ pieces

3. Window casing ________, ________′ pieces; ________, ________′ pieces

4. Window stool ________, ________′ pieces

5. Base molding _______ lineal feet

6. Crown molding _______ lineal feet

7. Chair rail _______ lineal feet

Estimate the interior trim materials needed for a 14′ 10″ × 17′ 10″ room with two 3′ 0″ × 4′ 6″ windows, one 6′ 0″ × 6′ 8″ trimmed archway, and one 6′ 0″ × 6′ 8″ exterior door. Show all of your work.

8. Exterior door casing _______, 7′ pieces

9. Interior door casing _______, 7′ pieces

Name ______________________ **Date** ____________ **Class** ____________

10. Window casing ________, ________′ pieces; ________, ________′ pieces

11. Window stool ________, ________′ pieces

12. Base molding ________ lineal feet

13. Crown molding ________ lineal feet

14. Chair rail ________ lineal feet

Estimate the interior trim materials needed for an 18′ 6″ × 23′ 4″ room with two 3′ 0″ × 5′ 0″ windows, one 6′ 0″ × 5′ 0″ window, one 4′ 0″ × 6′ 8″ trimmed archway, and one 8′ 0″ × 7′ 0″ exterior door. Show all of your work.

15. Exterior door casing ______, ______′ pieces; ______, ______′ pieces

16. Interior door casing ______, ______′ pieces; ______, ______′ pieces

17. Window casing ______, ______′ pieces; ______, ______′ pieces; ______, ______′ pieces

18. Window stool ______, ______′ pieces

19. Base molding ______ lineal feet

Name ______________________________ **Date** ____________ **Class** ____________

20. Crown molding ________ lineal feet

21. Chair rail ________ lineal feet

Work Space/Notes

ACTIVITY 10

Estimating Stair Frame Materials

Objective

After studying this section, you will be able to:

- Estimate stair frame materials.

Estimating stair materials occurs in two stages of the construction process. The first stage is when the stairs are constructed and finished off with risers and treads for a basement, attic, or exterior stairs. The first stage can also be when the main stairs of the house are framed and fitted with temporary treads for access during construction. Once the interior finish begins, these interior stairs will undergo a second stage. At this stage, the stairs will be trimmed with custom hardwood risers, treads, skirt boards, and a balustrade, which are purchased specifically for that staircase. This section will explain how to calculate the necessary materials needed during the first stage of stair construction with risers and treads.

Before any stair frame estimates can begin, the following information must be calculated first, regarding the future staircase:

- Stringer length
- Unit rise and run
- Number of risers and treads
- Length of the risers and treads

Example 22-12

The following image depicts a stairwell that will house the future stairs. Before any stair frame estimates can begin, the following information must be calculated first:

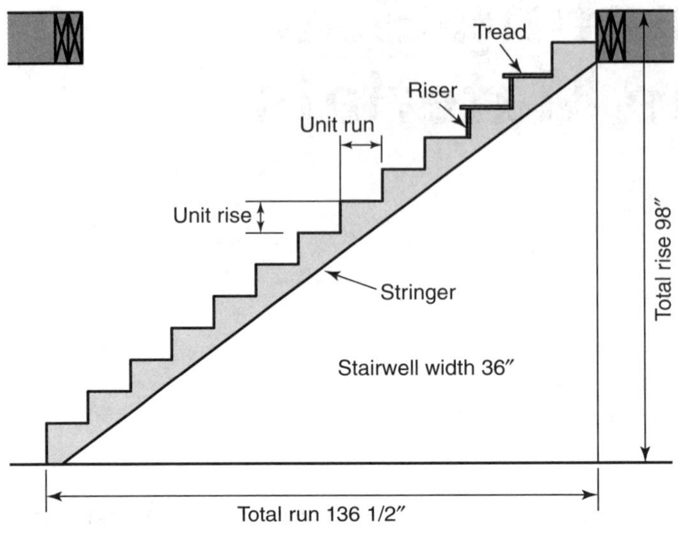

Goodheart-Willcox Publisher

1. Calculate stringer length. The length of the stringer can be determined using the Pythagorean theorem by inserting the total rise and total run into the equation (see Unit 21, *Right Angles*).

$$a^2 + b^2 = c^2$$

$$98^2 + 136.5^2 = 14'$$

Round up to the next even foot to account for angles and cutting.

16′ stringer length

2. Calculate unit rise and run and determine the number of risers and treads.

Determine total rise: 98″ inches

Divide total rise by 7 and drop any decimal:

$$98 \div 7 = 14 \text{ risers}$$

Divide total rise by number of risers:

$$98 \div 14 = 7'' \text{ riser}$$

Subtract one from number of risers:

$$14 - 1 = 13 \text{ treads}$$

Divide total run by number of treads:

$$136.5'' \div 13 = 10\ 1/2'' \text{ treads}$$

3. Length of the risers and treads. Always determined by the stairwell width. In this example, the stairwell width is 36″.

(Continued)

Using the calculations, the material estimates for stringers, risers, and treads can now be determined. Treads and riser board lengths are calculated to ensure the least amount of waste.

Material	Length	Quantity
2 × 12 Stringers	16′ long	3 pieces (per code)
1 × 8 Riser boards: 14 Risers each 3′ long	12′ long 6′ long	3 pieces 1 piece
2 × 12 Tread boards: 13 treads each 3′ long	12′ long 16′ long	2 pieces 1 piece

Goodheart-Willcox Publisher

Work Space/Notes

Name ______________________ Date ____________ Class ____________

ACTIVITY 10

Estimating Stair Frame Materials

Create a stair frame estimate. Use the following figure to solve problems 1–10.

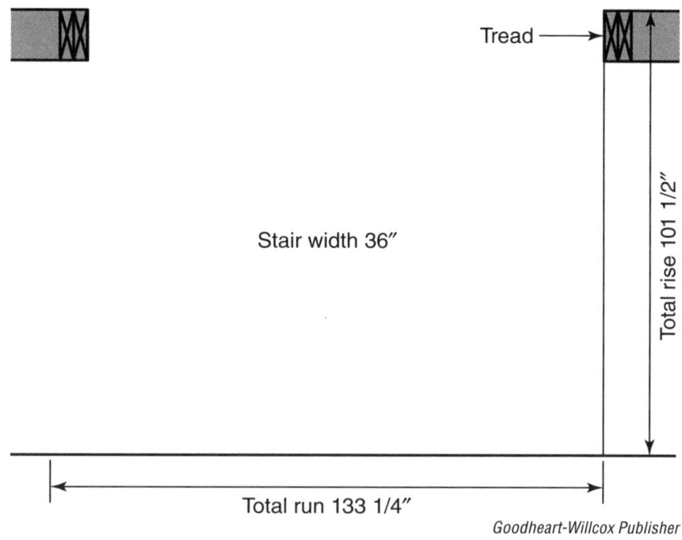

Goodheart-Willcox Publisher

1. Stringer length ________

2. Unit rise ________

3. Unit run ________

4. Number of risers ________

5. Number of treads ________

6. Riser length ________

7. Tread length ________

Name ______________________ **Date** ____________ **Class** ____________

Material Order:	Quantity	Length
8. 2 × 12 Stringers	______	______
9. 1 × 8 Riser boards:	______	______
	______	______
10. 2 × 12 Tread boards	______	______
	______	______

Create a stair frame estimate. Use the following figure to solve problems 11–20.

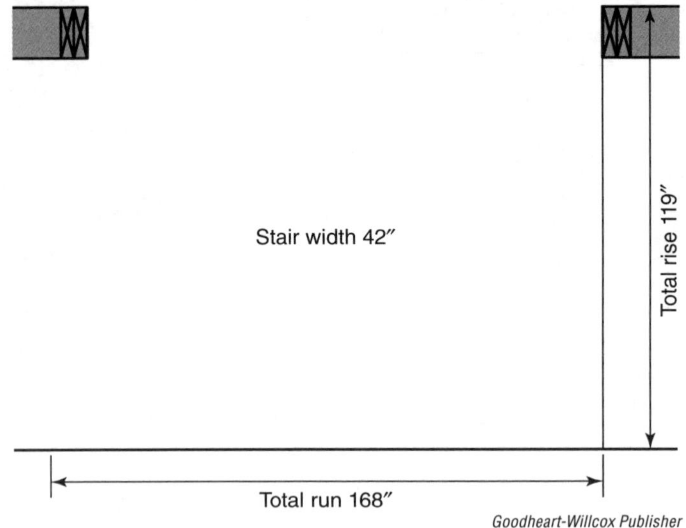

Goodheart-Willcox Publisher

11. Stringer length ________

12. Unit rise ________

13. Unit run ________

14. Number of risers ________

Name ______________________ **Date** __________ **Class** __________

15. Number of treads ________

16. Riser length ________

17. Tread length ________

Material Order:	**Quantity**	**Length**
18. 2 × 12 Stringers	________	________
19. 1 × 8 Riser boards	________	________
	________	________
20. 2 × 12 Tread boards	________	________

Work Space/Notes

APPENDIX A

Construction Diagrams and Terms

Foundation and Floor Frame

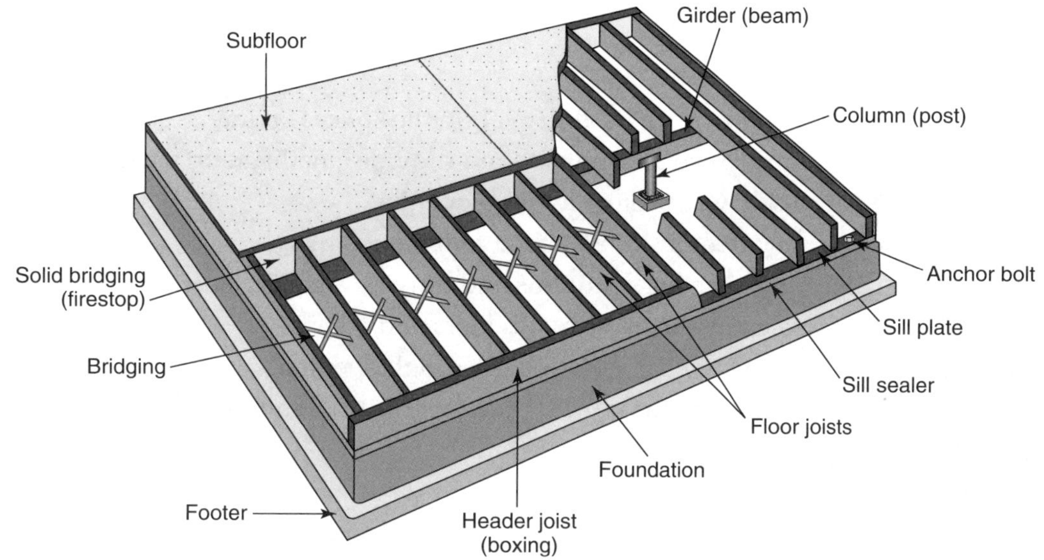

Goodheart-Willcox Publisher

anchor bolt. A metal fastener in the concrete that secures the sill plate to the foundation.
beam (girder). Engineered or built-up beam that supports the inner ends of the joists.
bridging. Diagonal bracing installed between floor joists to distribute the load on the floor.
column (post). A wood or steel vertical member used to support a girder.
floor joist. A horizontal framing member that provides support for the floor.
footer. A reinforced concrete slab that supports the foundation and prevents settling.
foundation. A concrete or block wall that supports the structure erected above it.
header joist (boxing). Joist material installed around the perimeter of the floor.
sill plate. The first horizontal wood member attached to the foundation with the anchor bolts.
sill sealer. Material installed between the foundation and sill plate to prevent air infiltration.
solid bridging (firestop). Solid blocks above a beam to prevent fires from spreading within a cavity.
subfloor. Sheathing material installed over the floor joists that is a base for the finished floor.

Wall Frame

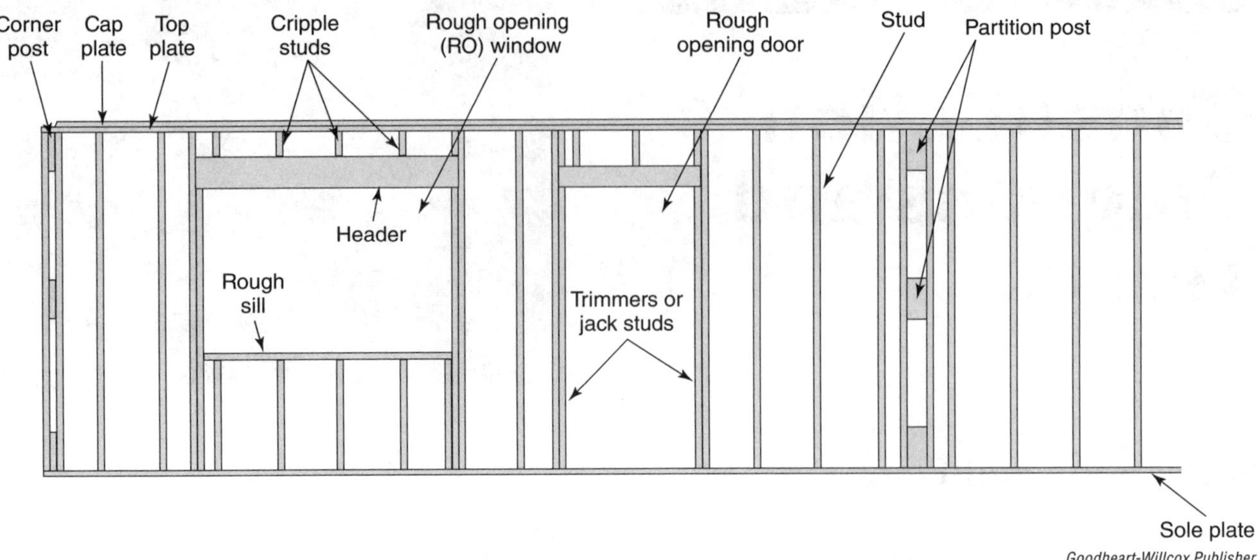

Goodheart-Willcox Publisher

cap plate. The second horizontal top plate used to overlap the splices on intersecting walls.
corner post. Used on outside corners to attach walls and provide a nailing surface for drywall.
cripple studs. Shortened studs above or below a rough opening.
header. A horizontal support member installed above rough openings.
partition post. Used to attach intersecting partitions and provide a nailing surface for drywall.
rough opening. Openings framed in a wall to install doors and windows.
rough sill. The horizontal bottom member of a window rough opening.
sole plate. The horizontal bottom plate that attaches the wall to the subfloor.
stud. The vertical on-center framing members of a wall.
top plate. The first horizontal top plate.
trimmer (jack stud). Studs located on the inside of a rough opening that support the header.

Roof Frame

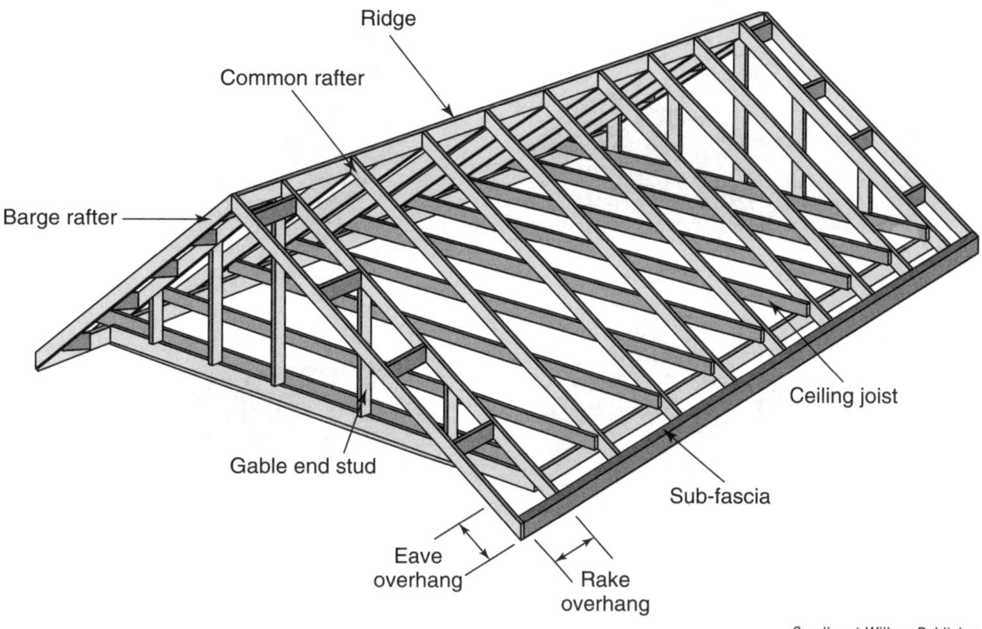

Goodheart-Willcox Publisher

barge rafter. The end common rafters that are part of the rake overhang of a gable roof.
ceiling joist. Horizontal joist that attaches to the top plates and ties the exterior walls together.
common rafter. The on-center roof member that runs from the top plate to the ridge.
eave overhang. Horizontal overhang consisting of rafter tails and sub-fascia.
gable end stud. Vertical studs in the gable end that are notched around the common rafters.
rake overhang. Sloped overhang that is made up of barge rafters and blocking.
ridge. Horizontal member that connects the upper end of the common and barge rafters.
sub-fascia. Horizontal board attached to the rafter tails that is a base for the finish fascia.

Roof Types and Terminology

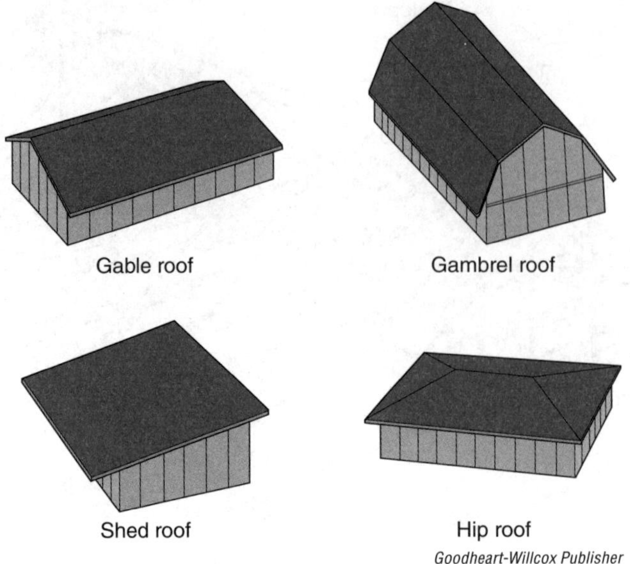

Goodheart-Willcox Publisher

gable roof. A roof consisting of a single ridge with slopes in both directions that is made entirely of common rafters.

gambrel roof. A variation of the gable roof, where each slope is broken at mid-span. Commonly used to create additional floor space in a one and a half story building.

hip roof. A roof that rises from all four sides of a building.

shed roof. A single-slope roof.

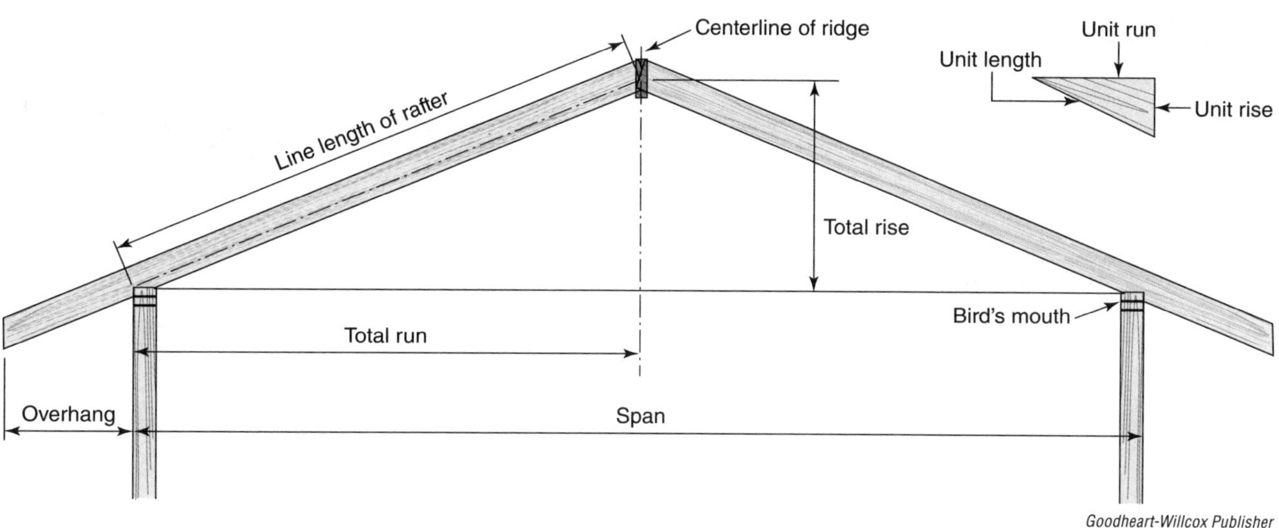

Goodheart-Willcox Publisher

bird's mouth. A notch cut on the lower end of a rafter where it is fastened to the wall.

line length of rafter. Length of the rafter from the outside edge of the plate to the centerline of the ridge.

overhang. Horizontal distance the rafter tail projects away from the wall.

span. The overall horizontal width of the building covered by the roof.

total rise. The vertical distance the roof rises from the top plate to the ridge.

total run. One-half of the span or the horizontal distance the rafter covers.

unit length. The distance the rafter travels for every unit of run.

unit rise. The height the rafter rises for every unit of run.

unit run. The standard horizontal distance of 12″ used for all common rafters.

Stair Frame

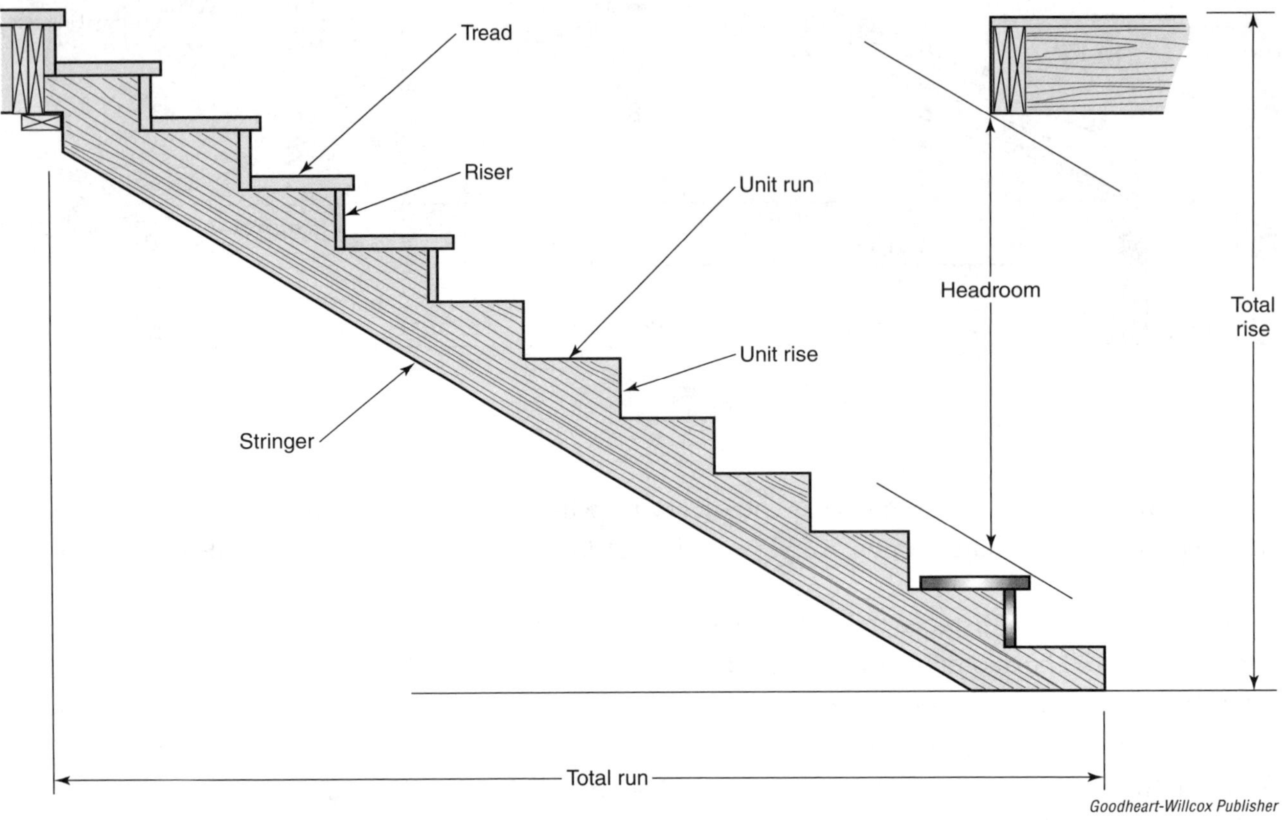

Goodheart-Willcox Publisher

headroom. The minimum vertical distance permitted between the stairs and ceiling.
riser. The finished board installed on each unit rise.
stringer. The main board that supports the risers and treads.
total rise. The vertical distance between the two finished floors.
total run. The horizontal distance the stairs cover.
tread. The finished board installed on each unit run.
unit rise. The vertical distance between each step.
unit run. The horizontal distance between each riser.

Interior Finish

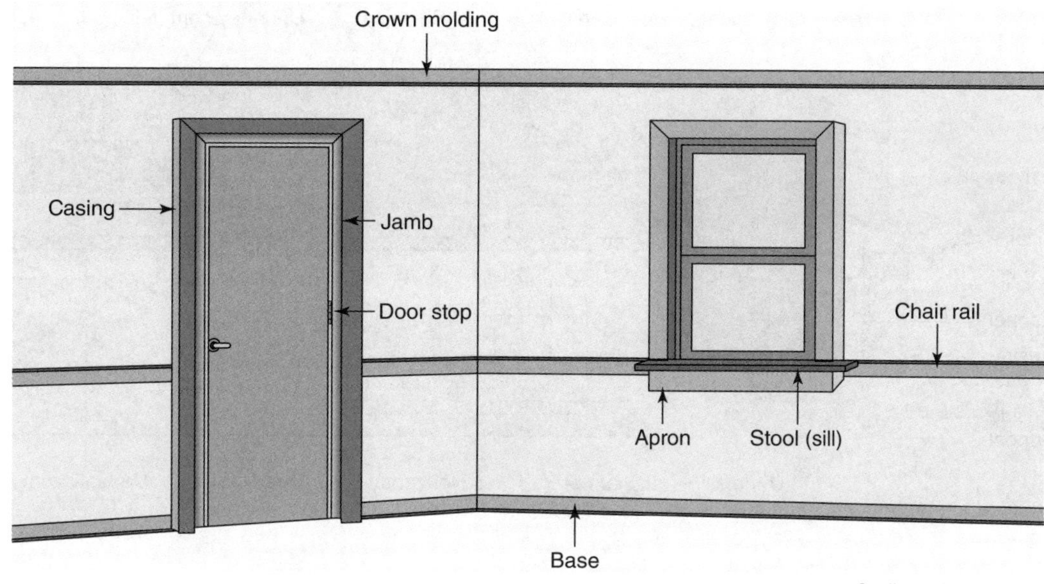

Goodheart-Willcox Publisher

apron. A piece of casing material with returned ends to support the stool.
base. Molding applied at the intersection of the finished floor and wall.
casing. Molding installed on the side and top of doors and windows.
chair rail. Horizontal trim used as a top cap for wainscoting.
crown molding. Horizontal molding installed where walls and ceilings intersect.
door stop. Molding applied to doorjambs to stop the door when it is closed.
jamb. The vertical side (side jamb) and horizontal top (head jamb) of doors and windows.
stool (sill). The bottom horizontal finished trim of the window.

Exterior Finish

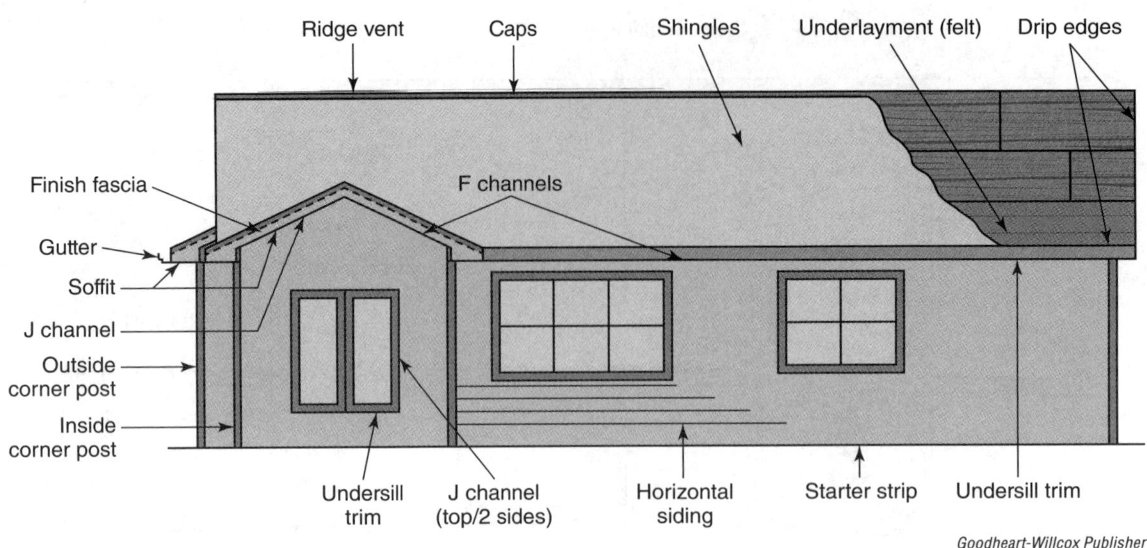

Goodheart-Willcox Publisher

cap. Shingle tabs applied to the top of the roof after both sides have been shingled.

drip edge. Aluminum edging material applied around the perimeter of the roof before installing shingles.

F-channel. F-shaped trim used under eaves and rakes to secure soffit against the wall.

finish fascia. Wood or aluminum finishing material applied over the sub-fascia.

gutter. An exterior trough along the eaves to collect and divert water from rain or snow.

horizontal siding. Wood, aluminum, or vinyl finish material applied horizontally on the walls.

inside corner post. Vertical trim piece designed to finish the inside corner.

J-channel. Vinyl or aluminum material used around windows, doors, and rakes.

outside corner post. Vertical trim piece designed to finish the outside corner.

ridge vent. A material installed under the caps that allows for ventilation at the ridge.

shingles. Asphalt strip material with a mineral granular surface designed to protect the roof.

soffit. Perforated (for ventilation) or solid material to finish the underside of rakes and eaves.

starter strip. Horizontal strip installed at the perimeter of the building to attach the siding.

underlayment (felt). Felt paper saturated with tar that protects the roof prior to shingling.

undersill trim. Component installed under windows and eaves to fasten the siding.

APPENDIX B

References

Common Decimal Equivalents

Fraction	Decimal	Fraction	Decimal	Fraction	Decimal	Fraction	Decimal
1/64	0.015625	17/64	0.265625	33/64	0.515625	49/64	0.765625
1/32	0.03125	9/32	0.28125	17/32	0.53125	25/32	0.78125
3/64	0.046875	19/64	0.296875	35/64	0.546875	51/64	0.796875
1/16	0.0625	5/16	0.3125	9/16	0.5625	13/16	0.8125
5/64	0.078125	21/64	0.328125	37/64	0.578125	53/64	0.828125
3/32	0.09375	11/32	0.34375	19/32	0.59375	27/32	0.84375
7/64	0.109375	23/64	0.359375	39/64	0.609375	55/64	0.859375
1/8	0.125	3/8	0.375	5/8	0.625	7/8	0.875
9/64	0.140625	25/64	0.390625	41/64	0.640625	57/64	0.890625
5/32	0.15625	13/32	0.40625	21/32	0.65625	29/32	0.90625
11/64	0.171875	27/64	0.421875	43/64	0.671875	59/64	0.921875
3/16	0.1875	7/16	0.4375	11/16	0.6875	15/16	0.9375
13/64	0.203125	29/64	0.453125	45/64	0.703125	61/64	0.953125
7/32	0.21875	15/32	0.46875	23/32	0.71875	31/32	0.96875
15/64	0.234375	31/64	0.484375	47/64	0.734375	63/64	0.984375
1/4	0.25	1/2	0.5	3/4	0.75	1	1.0

Goodheart-Willcox Publisher

Inches Converted to Decimals of Feet

Inches	Decimal of a Foot	Inches	Decimal of a Foot	Inches	Decimal of a Foot
1/8	0.01042	3	0.25000	6	0.50000
1/4	0.02083	3 1/8	0.26042	6 1/4	0.52083
3/8	0.03125	3 1/4	0.27083	6 1/2	0.54167
1/2	0.04167	3 3/8	0.28125	6 3/4	0.56250
5/8	0.05208	3 1/2	0.29167	7	0.58333
3/4	0.06250	3 5/8	0.30208	7 1/4	0.60w417
7/8	0.07291	3 3/4	0.31250	7 1/2	0.62500
1	0.08333	3 7/8	0.32292	7 3/4	0.64583
1 1/8	0.09375	4	0.33333	8	0.66666
1 1/4	0.10417	4 1/8	0.34375	8 1/4	0.68750
1 3/8	0.11458	4 1/4	0.35417	8 1/2	0.70833
1 1/2	0.12500	4 3/8	0.36458	8 3/4	0.72917
1 5/8	0.13542	4 1/2	0.37500	9	0.75000
1 3/4	0.14583	4 5/8	0.38542	9 1/4	0.77083
1 7/8	0.15625	4 3/4	0.39583	9 1/2	0.79167
2	0.16666	4 7/8	0.40625	9 3/4	0.81250
2 1/8	0.17708	5	0.41667	10	0.83333
2 1/4	0.18750	5 1/8	0.42708	10 1/4	0.85417
2 3/8	0.19792	5 1/4	0.43750	10 1/2	0.87500
2 1/2	0.20833	5 3/8	0.44792	10 3/4	0.89583
2 5/8	0.21875	5 1/2	0.45833	11	0.91667
2 3/4	0.22917	5 5/8	0.46875	11 1/4	0.93750
2 7/8	0.23959	5 3/4	0.47917	11 1/2	0.95833
		5 7/8	0.48958	11 3/4	0.97917
				12	1.00000

Goodheart-Willcox Publisher

US Customary Conversion Charts

Distance
12 inches = 1 foot
3 feet = 1 yard
36 inches = 1 yard
1,760 yards = 1 mile
5,280 feet = 1 mile
Area
144 sq in = 1 sq ft
9 sq ft = 1 sq yd
14,520 sq yd = 1 acre
43,560 sq ft = 1 acre
Volume
1,728 cu in = 1 cu ft
27 cu ft = 1 cu yd
Weight
16 ounces = 1 pound
2,000 pounds = 1 ton
Liquid Volume
8 fluid ounces = 1 cup
2 cups = 1 pint
2 pints = 1 quart
4 quarts = 1 gallon
Time
60 seconds = 1 minute
60 minutes = 1 hour
24 hours = 1 day
7 days = 1 week
365 days = 1 year
52 weeks = 1 year
12 months = 1 year

Goodheart-Willcox Publisher

Nominal Dressed Lumber Sizes

Nominal Size	Dressed Size	Nominal Size	Dressed Size
1 × 2	3/4″ × 1 1/2″	2 × 8	1 1/2″ × 7 1/4″
1 × 3	3/4″ × 2 1/2″	2 × 10	1 1/2″ × 9 1/4″
1 × 4	3/4″ × 3 1/2″	2 × 12	1 1/2″ × 11 1/4″
1 × 6	3/4″ × 5 1/2″	4 × 4	3 1/2″ × 3 1/2″
1 × 8	3/4″ × 7 1/4″	4 × 6	3 1/2″ × 5 1/2″
1 × 10	3/4″ × 9 1/4″	4 × 8	3 1/2″ × 7 1/4″
1 × 12	3/4″ × 11 1/4″	4 × 10	3 1/2″ × 9 1/4″
2 × 2	1 1/2″ × 1 1/2″	4 × 12	3 1/2″ × 11 1/4″
2 × 3	1 1/2″ × 2 1/2″	6 × 6	5 1/2″ × 5 1/2″
2 × 4	1 1/2″ × 3 1/2″	6 × 8	5 1/2″ × 7 1/4″
2 × 6	1 1/2″ × 5 1/2″		

Goodheart-Willcox Publisher

Summary of Formulas

Square Perimeter

$P = s + s + s + s$

$P = 4 \times s$

Square Area

$A = s \times s$

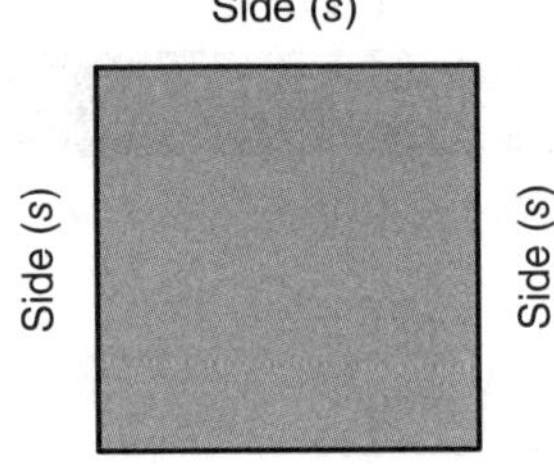

Rectangle Perimeter

$P = \text{length}(l) + \text{width}(w) + \text{length}(l) + \text{width}(w)$

$P = (2 \times l) + (2 \times w)$

Rectangle Area

$A = l \times w$

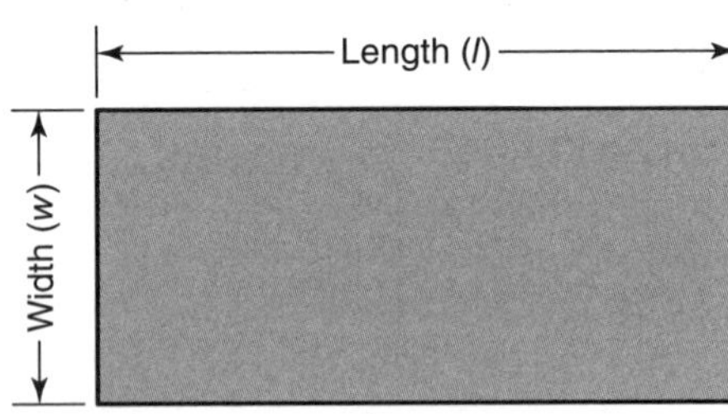

Triangle Area

$A = 1/2\,(b \times h)$

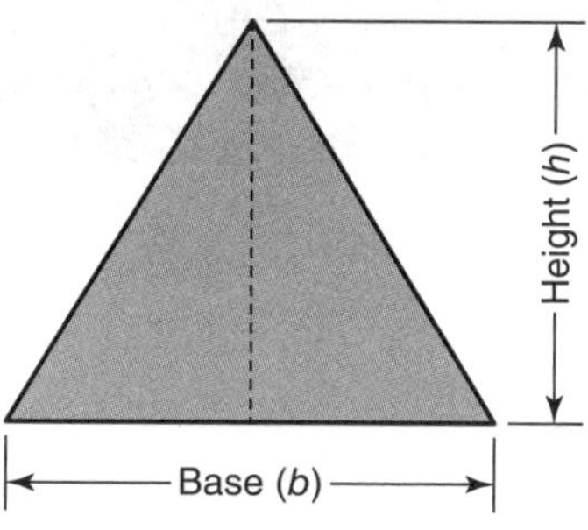

Cube Volume

$V = s \times s \times s$

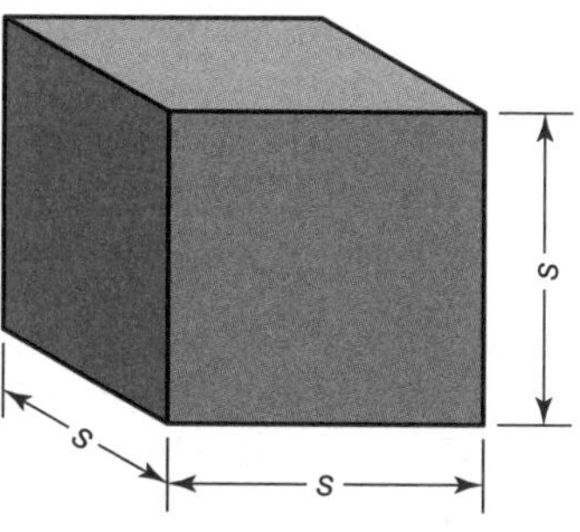

Rectangular Solid Volume

$V = l \times w \times h$

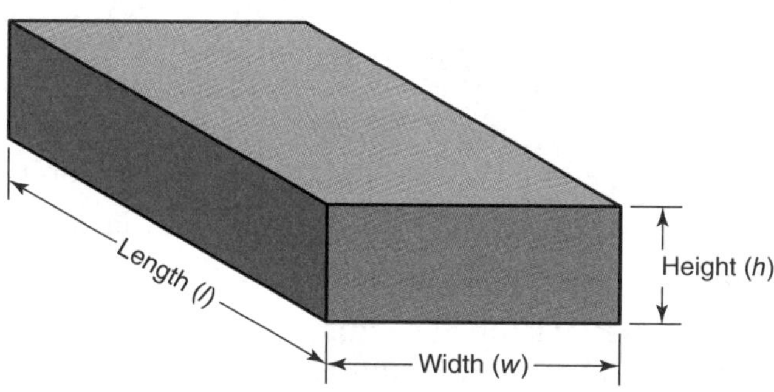

Pythagorean Theorem

$a^2 + b^2 = c^2$

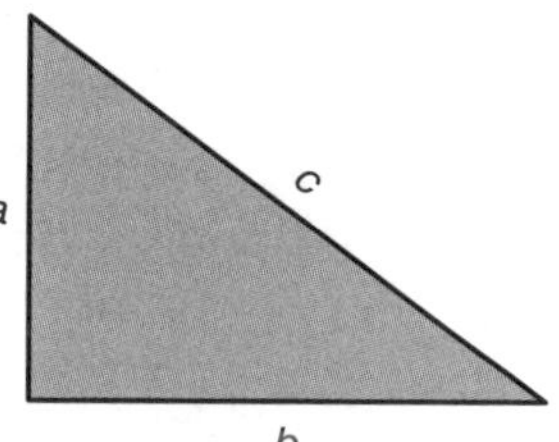

Board Feet

$$\text{Board feet} = \frac{t\text{ (inches)} \times w\text{ (inches)} \times l\text{ (feet)}}{12}$$

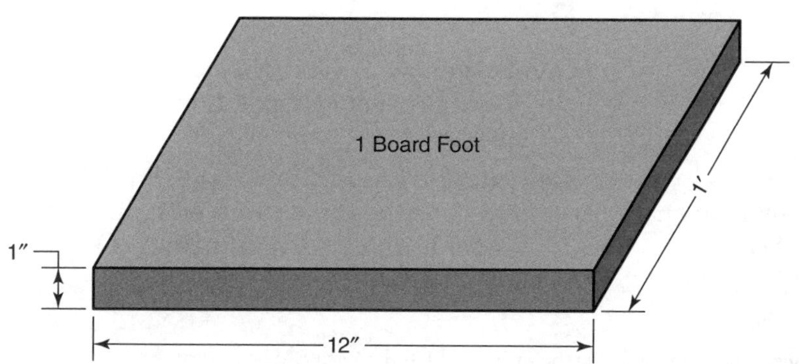

GLOSSARY

3-4-5 method. A method to establish a right triangle using 3, 4, and 5 units as lengths of the sides. (21)

A

addition. The process of combining two or more numbers to obtain a sum. (2)

architect's scale. A specialized device consisting of eleven scales used to facilitate measuring on architectural drawings. (15)

area. The amount of surface within a polygon, measured in square units. (17)

B

board foot. A board whose volume is 144 cubic inches. (19)

borrowing. The process of taking one unit from a place value to the left of the column being subtracted. (3)

C

cancel. A method used to simplify fraction multiplication problems. Numbers that can be divided evenly into the numerator and denominator are simplified. (9)

carrying. In a mathematical process, moving the second digit of a sum to the next place value. (2)

common denominator. A denominator value that is the same in a set of fractions. (6)

cube. A three-dimensional square that has equal dimensions along its length, width, and height. (18)

D

decimal. A number that is less than a whole and is written with a decimal point. (11)

decimal number system. The numbering system commonly used today which uses 10 as its base unit and consists of ten different digits (0, 1, 2, 3, 4, 5, 6, 7, 8, and 9). Also known as the Arabic number system. (1)

decimal point. In a decimal number, a point that separates the whole number portion from the decimal portion. (11)

denominate number. An expression consisting of a numeric value and a unit of measure. (1)

denominator. The bottom number in a common fraction. (6)

difference. The number resulting from a subtraction problem. (3)

digit. A symbol (0, 1, 2, 3, 4, 5, 6, 7, 8, or 9) that is used to create numbers. (1)

dividend. The number in a division equation that is divided into smaller parts by the divisor. (5)

division. The process of separating a quantity into a specific number of equally sized portions. (5)

division sign (÷). A symbol placed between numbers to indicate the division process. (5)

divisor. The number by which the dividend is being divided. (5)

dressed size. The size a board actually measures after milling and planing. (19)

E

equals sign (=). A symbol placed at the end of a mathematical operation to indicate an answer. (2)

equivalent fraction. Two or more fractions that have the same value with different numerators and denominators. (6)

Note. The number in parentheses following each definition indicates the unit in which the term can be found.

F

fraction. Two numbers separated by a division line used to express the part of a whole. (6)

fraction bar. The line that separates the numerator and the denominator in a fraction. The fraction bar has the same meaning as the ÷ sign. (6)

G

graduations. A series of standard lines on a rule used for measuring distance. (15)

H

higher terms. The process of making the terms of a fraction higher without changing the fraction value. (6)

hypotenuse. The longest side of a right triangle. (21)

I

improper fraction. A fraction whose numerator is greater than the denominator. (6)

L

lineal measurement. A measurement of the distance between two points. (15)

lowest common denominator (LCD). The smallest shared multiple of a set of denominators. (6)

lowest terms. A fraction is considered to be in its lowest term when it has the lowest possible numerator and denominator. (6)

M

minus sign (–). The symbol used to identify a subtraction operation. (3)

mixed number. A number that consists of a whole number combined with a fraction. (6)

multiplication. A method of repeat addition or adding the same number together a certain number of times. (4)

multiplication sign (×). The symbol used to identify a multiplication operation. (4)

N

nominal size. The size that boards are referred to, not their actual size. (19)

numerator. The top number in a common fraction. (6)

P

percentage. A part of a whole, representing a fraction of 100. (20)

percent symbol. The symbol used with a number to indicate a percentage. (20)

perimeter. The distance around the outside of a polygon. (16)

place value. The value placed on a digit based on its location within a number. (1)

plus sign (+). A symbol placed between numbers to indicate the addition process. (2)

polygon. An enclosed geometric figure, created by three or more straight lines. (16)

prime factors. All the prime numbers a larger number can be evenly divided into. (6)

prime number. A number that can only be divided evenly by 1 and itself. (6)

product. A quantity obtained by multiplying quantities together. (4)

proper fraction. A fraction whose numerator is less than the denominator. (6)

Pythagorean theorem. A theory that states the square of the hypotenuse is equal to the sum of the squares of the two shorter sides, $a^2 + b^2 = c^2$. (21)

Q

quotient. The number of times the dividend is divided by the divisor. (5)

R

reciprocal. A value used when dividing fractions, found by inverting the divisor's numerator and denominator. Also, the value to multiply a number by in order to have a product of 1. (10)

rectangle. A four-sided polygon whose opposite sides are both parallel and equal in length with 90° corners. (16)

rectangular solid. A three-dimensional rectangle with 6 faces. (18)

reducing. A process used to convert a fraction to its lowest term. (6)

remainder. A number that remains after the entire dividend has been divided and is attached to the quotient with the letter "r." (5)

right angle. A 90° angle made up of two perpendicular lines. (21)

right triangle. A triangle that contains one 90° angle. (21)

rough cut. A board that has been milled, but has not been planed to a dressed dimension. (19)

rounding. The process of increasing or decreasing the value of the number to a specific place value. (1)

S

scale. 1—A measuring device with standard graduations. (15) 2—A ratio between the actual size of a project and the size it is actually drawn. (15)

square. 1—Equals 100 square foot. 2—A four-sided polygon that has equal measurements on all four sides and four 90° corners. (16) 3—A number multiplied by itself. (21)

square root. A number that produces a specified quantity when multiplied by itself. (21)

subtraction. The process of removing units to find the difference between numbers. (3)

sum. The combined total of two or more numbers using the mathematical process of addition. (2)

T

triangle. An enclosed polygon formed by three straight lines. (17)

U

US Customary system. The standard system of measure used in the United States, consisting of units such as the inch, foot, yard, and the mile. (15)

V

volume. The amount of space within a three-dimensional object. (18)

W

whole number. A number that stands alone as a unit and does not include a fraction or a decimal. Can contain one or multiple digits. (1)

INDEX

S

T

U

V

W

Y

ANSWERS TO ODD-NUMBERED QUESTIONS

UNIT 1
Basic Principles of Whole Numbers

1. 5
3. 2
5. 4
7. Hundred Thousands
9. Tens
11. 170,000
13. 93,830
15. 8,500
17. 900
19. 31,720,000
21. 93″
23. 16′, 14′
25. 7-acre

UNIT 2
Adding Whole Numbers

1. 97
3. 143
5. 612
7. 2,572
9. 10,347
11. 200″
13. 2,030 yd
15. $8,630
17. Base = 342 lineal feet
    Casing = 837 lineal feet
    Crown = 256 lineal feet
    Chair Rail = 94 lineal feet
19. $10,942
21. 4,540 sq ft
23. 768″
25. One-Gang Squares = 49 boxes
    Two-Gang Squares = 15 boxes
    Round Ceiling = 9 boxes
27. 152 lineal feet
29. 8′ 11″

UNIT 3
Subtracting Whole Numbers

1. 63
3. 4,118
5. 97
7. 105
9. 290 sq ft
11. 1,011 bd ft
13. 672,462
15. $6,155
17. 24 sheets
19. Distance D = 7′
21. Distance B = 9′
23. 15 man-hours
25. 1,675 sq ft

UNIT 4
Multiplying Whole Numbers

1. 35
3. 16
5. 21
7. 192
9. 392
11. 432

13. 588
15. 1,998
17. 27,360
19. 407,925
21. 432,024
23. 963,736
25. 14,748,656
27. $1,440
29. 120 feet
31. 1,152 sq ft
33. 504 gang plates
35. $3,748

UNIT 5 Dividing Whole Numbers

1. 5
3. 3
5. 5
7. 86
9. 621 r7
11. 841 r10
13. 126 r92
15. 63
17. 123
19. 177 r244
21. 6 hours
23. 4 pieces
25. 37 trusses
27. 21 boards
29. 14 squares
31. 196 ceiling tiles
33. 10″
35. 45 hours

SECTION 1 Exam

1. 9
3. Hundred thousands
5. 67″
7. 637 sq ft
9. 55 sq ft
11. 4,712 bd ft
13. 714
15. 1,830,675
17. 52 r19
19. 81 r193
21. 196 hours
23. 79″
25. 64

UNIT 6 Basic Principles of Fractions

1. 4
3. 40
5. 56
7. 22
9. 5
11. 1/9
13. 3/8
15. 3/8
17. 11/27
19. 3/80
21. 1 3/4
23. 7 1/4
25. 17 1/3
27. 1 11/21
29. 2 13/16
31. 13/8
33. 115/16
35. 30/8
37. 60
39. 72

UNIT 7 Adding Fractions

1. 7/8
3. 10/15 = 2/3
5. 19/21
7. 17/60
9. 5/14

11. 187/96 = 1 91/96
13. 20 7/30
15. 162 9/32
17. 17/16″ = 1 1/16″
19. 11 15/16″
21. 61 3/16″
23. 9 1/2″
25. 15 2/3 cu yd
27. 3′ 3 3/4″
29. 11 3/4″
31. 5 3/16″
33. 2′ 8 1/2″
35. 31 1/8″

UNIT 8 Subtracting Fractions

1. 1/2
3. 3/16
5. 5/16
7. 27/64
9. 7 2/9
11. 26 5/6
13. 20 31/32
15. 5 3/8″
17. 44 7/12 sq ft
19. 13/16″
21. 80 3/4″
23. 1 7/8″
25. 2 1/4″
27. 8 9/16″
29. 11 1/4″

UNIT 9 Multiplying Fractions

1. 3/16
3. 3/40
5. 3/152
7. 36 9/56
9. 218 1/2
11. 55 85/128
13. 118 1/8
15. 23 29/32
17. 373 31/128
19. 104 1/2″
21. 170″
23. 150 3/8″
25. 45 1/4″
27. 45 3/8″
29. 46′

UNIT 10 Dividing Fractions

1. 5/9
3. 2/3
5. 1 1/24
7. 10 2/3
9. 5 13/23
11. 4 7/8
13. 2 4/17
15. 2 13/30
17. 29/73
19. 7
21. 10 1/2″
23. 34
25. 3 1/2″
27. 35
29. $24/hour

SECTION 2 Exam

1. 45
3. 5 3/7
5. 72
7. 31 3/8
9. 3/32
11. 15 11/16″
13. 33
15. 3 5/6
17. 3 3/4

19. 108 7/8″
21. 2 5/8″
23. 126 1/2″
25. 41 rows
27. 3.72′
29. $672.50

UNIT 11
Basic Principles of Decimals

1. 6.739
3. 0.66284
5. 0.2314
7. 73/200
9. 96/125
11. 2 15/16
13. 0.3
15. 0.02
17. 0.94
19. 0.197
21. 0.863
23. 114 7/8″
25. 5/16″

UNIT 12
Adding and Subtracting Decimals

1. 12.58
3. 180.504
5. 4.8
7. 15.07
9. 7.7
11. 555.541
13. $500.15
15. 1.46875″
17. $1172.45
19. $6071.62
21. 15.75″
23. $28.25
25. 2.345′

UNIT 13
Multiplying Decimals

1. 32
3. 354
5. 434.07
7. 0.2808
9. 0.1136853
11. 3.96185
13. 177 bd ft
15. 103.25″
17. 27 hours
19. 206.25″
21. 51″
23. $5940.38
25. 3,208.8 lbs

UNIT 14
Dividing Decimals

1. 172.5
3. 3.265
5. 8.6
7. 0.053
9. 10
11. 6.807
13. 0.128
15. 6.596
17. 0.45
19. 0.8
21. 0.4
23. $2.23 per bd ft
25. 12.5 hours
27. 14.29 shingles
29. 15 risers
31. 56
33. 45

35. Stringer 1—7 risers
    Stringer 2—8 risers
37. 107.125″

SECTION 3 Exam

1. 0.5
3. 0.005
5. 573.67
7. 112.09
9. 4.5084
11. 1.9604112
13. 8.25
15. 0.1875
17. 0.6875
19. 127 1/8″ × 98 5/8″
21. $342.77
23. $281.60
25. 24

UNIT 15 Linear Measurement

1 11/16″
3. 2 1/8″
5. 3 1/2″
7. 2′ 7 1/2″
9. 3′ 3/8″
11. 3′ 2 1/8″
13. 6′ 3″
15. 2′ 10″
17. 34″
19. 152″
21. 52 1/2″
23. 3′ 5″
25. 4′ 9 1/16″
27. 2.917′
29. 7.417′
31. 11.75′
33. 11′ 3 7/16″
35. 10′ 10 3/4″

UNIT 16 Perimeter Measurement

1. 96′
3. 430′
5. 54′
7. 17 pieces
9. 11 sill plates
11. 12 posts
13. 22 tack strips
15. 158′
17. 141′ 4″
19. 13 pieces
21. 67′ 6″
23. 14′ 6″
25. 50′

UNIT 17 Area Measurement

1. 81 sq in
3. 96 sq in
5. 35 sq ft
7. 306 sq ft
9. 341.25 sq ft
11. 141 sq ft
13. 174 sq ft
15. 2.53
17. 806.4
19. 12
21. 3144 sq ft
23. 23.33 sheets
25. 16.02 pieces
27. 4 gallons
29. 10 bags

UNIT 18 Volume Measurement

1. 343 cu in
3. 1,838.27 cu ft

5. 360 cu in
7. 1.56 cu yd
9. 4.14 cu yd
11. 5.03
 0.79
13. 2
 0.3
15. 4
 0.67
17. 243
19. 9
21. 256,608
23. 0.67 cu yd
25. 2.85 cu yd
27. 433.33 cu yd
29. 18.23 cu yd

UNIT 19 Board Foot Measurement

1. 9″
3. 20 1/2″
5. 6″
7. 23
9. 28″
11. 93.333
13. 234.667
15. 616
17. 704
19. 80
21. 96
23. 161.333
25. $272.64

SECTION 4 Exam

1. 55″
3. 9′ 10 1/4″
5. 18′ 9 3/4″
7. 181′
9. 138 sq ft
11. 4.5 sq ft
13. 315 cu in
15. 576 bd ft
17. 648 cu ft
19. 9.25 plates
21. 59.5′
23. 252 bd ft
25. 121.41 bd ft

UNIT 20 Percentages

1. 73/100
3. 1 13/20
5. 1/320
7. 49/150
9. 0.23
11. 0.245
13. 0.00138
15. 3.19
17. 73%
19. 4.6%
21. 494%
23. 9602%
25. 70%
27. 162.5%
29. 415%
31. 12
33. 11.11%
35. 30
37. 18.4%
39. $27,580

UNIT 21 Right Angles

1. 21′ 7 5/8″
3. 14′ 11 5/8″
5. 23′ 2 7/8″
7. 33′ 6 1/8″
9. 22′ 3 3/8″
11. 22′ 4 11/16″
13. 9′ 2″

15. 48′
17. 6′
19. 75′

15. 144
17. 172
19. 11
21. 58
23. 112

SECTION 5 Exam

1. 11/20
3. 0.85
5. 98%
7. 62.5%
9. 28.8
11. 30
13. 8′ 4″
15. 27′ 9 7/16″
17. 10′ 6″
19. 21′ 9 15/16″
21. 24′ 8 7/8″
23. $78,750
25. 16′ 2″

SECTION 6

Activity 1: Estimating Concrete for Slabs, Footers, and Walls

1. 2.8
3. 1.7
5. 3.9
7. 9.3
9. 36.5

Activity 2: Estimating Floor Frame Materials

1. 212
3. 14
5. 114
7. 222
9. 148
11. 10
13. 74

Activity 3: Estimating Wall Frame Materials

1. 9
3. 13
5. 47
7. 11
9. 14

Activity 4: Estimating Ceiling Frame Materials

1. 68
3. 54
5. 66
7. 66
9. 62

Activity 5: Estimating Roof Frame Materials

1. 156″
3. 3
5. 187.88″
7. 4
9. 164.45″
11. 3

Activity 6: Estimate Roof Finish Materials

1. 13
3. 9.74
5. 1512
7. 15
9. 14.28
11. 2953

Activity 7: Estimating Siding Materials

1. 5
3. 6
5. 4
7. 5
9. 4
11. 4
13. 2
15. 4

Activity 8: Estimating Insulation Materials

1. 17
3. 23
5. 4
7. 12
9. 18

Activity 9: Estimating Interior Trim Materials

1. 3
3. 2, 8′; 2, 11′
5. 62
7. 56
9. 6
11. 1, 8′
13. 66
15. 2, 8′; 1, 9′
17. 2, 7′; 2, 8′; 3, 11′
19. 72
21. 60

Unit 10: Estimate Stair Frame Materials

1. 14′
3. 10 1/4″
5. 13
7. 36″
9. 3, 12′
   1, 6′
11. 18′
13. 10 1/2″
15. 16
17. 42″
19. 4, 14′
   1, 4′